Scotland's Science

Stories of pioneering science, engineering
and medicine
(1550-1900)

John Mellis

Copyright © 2020 John Mellis
All rights reserved
ISBN-979 8 64584 962 7

Acknowledgements

I am grateful to the many people who helped in the conception and completion of the project, including of course all my family and friends, who showed great interest throughout. Dr. Roger Steele provided valuable comments on the completed manuscript. Dr. Paul 'Griff' Griffin of the Department of Physics at the University of Strathclyde pointed me to John Butt's history of that university and its Andersonian antecedents. Gavin MacDougall at Luath Press gave useful feedback on the proposal, which reminded me of the inspiring story of the Edinburgh Seven.

To Anthea, with great love and thanks for her forbearance and support

Contents

Foreword v

Part One: Out of the Dark Ages 7

1. Out of the Dark Ages 9
2. The Logs and Bones of 'Marvellous Merchiston': John Napier (1550-1617) 12
3. Scientist, freemason, soldier, spy - and founder of the Royal Society: Robert Moray (1609-1673) 16
4. The maritime pendulum clock: Alexander Bruce (1629-1681) 26
5. Royal botanist and physician: Robert Morison (1620-1683) 31
6. Telescopes and trigonometry: James Gregory (1638-1675) 35

Part Two: The Scottish Enlightenment 41

7. The Scottish Enlightenment 43
8. The world's youngest professor: Colin Maclaurin (1698-1746) 47
9. Hume's Fork and Humanism: David Hume (1711 – 1776) 53
10. Founding the science of economics: Adam Smith (1723-1790) 61
11. Anatomy, obstetrics and scientific surgery: William and John Hunter (1718-1793) 67

12 Lemons, Limeys, and Scots against scurvy: James Lind, Gilbert Blane and Lauchlan Rose (1716-1885) 75

13 Natural philosopher and natural educator: John Anderson (1726-1796) 84

14 The genesis of geology: James Hutton (1726-1797) 90

15 Shedding light on heat, and the 'discovery' of air: Joseph Black and Daniel Rutherford (1729-1819) 96

16 Steam power efficiency: James Watt (1736-1819) 102

17 Cerebral geniuses: Alexander Monro *secundus* and George Kellie (1733-1829) 113

18 The Colossus of roads: John Loudon McAdam (1756-1836) 118

19 Master of civil engineering: Thomas Telford (1757 – 1834) 125

20 Rainwear from raintown: Charles Macintosh (1766-1843) 132

21 Enlightenment lights: Robert Stevenson (1772- 1850) 138

22 Through a lens brightly: Robert Brown (1773-1858) 145

Part Three: The Imperial impulse 151

23 The Imperial impulse 152

24 The Queen of 19th century science: Mary Somerville (1780-1872) 155

25 Father of modern optics: David Brewster (1781-1868) 162

26 Pioneering polarizer: William Nicol (c.1770-1851) 170

27 The Stirling Engine: Robert Stirling (1790-1878) 174

28 Anatomy at the edge of legality: Robert Knox (1791-1862) 179

29 Geological time and the antiquity of Man: Charles Lyell (1797-1875) 188

30 Electric clocks and electric thoughts: Alexander Bain (1810-1877) 195

31 Diffusion and dialysis: Thomas Graham (1805-1869) 204

32 Great ships and solitary waves: John Scott Russell (1808-1882) 210

33 Light from oil: James 'Paraffin' Young (1811-1883) 218

34 Childbirth and anaesthesia: James Young Simpson (1811-1870) 228

35 Antiseptic surgery: Joseph Lister (1827-1912) 237

36 First female doctors of medicine: Elizabeth Garrett Anderson and the Edinburgh Seven (1836-1900) 247

37 Magneto- and electro-optics: John Kerr (1824-1907) 259

38 The Rankine Cycle: William John Macquorn Rankine (1820-1872) 263

39 Absolute temperature and Thermodynamics: William Thomson - Lord Kelvin (1824-1907) 269

40 The beautiful equations that changed the world: James Clerk Maxwell (1831 –1879) 284

41 Into the 20th century 301

42 Sources and further reading 302

Foreword

These stories richly deserve to be told, and of course to some extent many have been, individually, in biographies, encyclopaedias, and valedictory obituaries. As a collection, I think they are even more impressive. The thread that connects them is Scotland, a small nation on the northern outskirts of Europe which had no obvious credentials to emerge from her especially turbulent Middle Ages to be one of the world's foremost leaders in scientific thought and practice. Many theories have been advanced to explain this apparent anomaly. High rates of literacy were established through the proliferation of Church schools post-Reformation. The five ancient Scottish universities provided easy and early access to a broad-based higher education which avoided student specialisation at too young an age. Historic ties and alliances with France, the Low Countries and the German states encouraged Scottish scholars to study abroad and fostered an open, secular mindset. The colonial power of neighbouring England and her Industrial Revolution provided an ideal outlet for Scotland's practical sciences. There may be some truth in all of these assertions. Whatever the truth, a collection of stories, similar to that which follows, could equally be written for the scientists of France, Italy, Germany or England. And because scientific progress depends above all on communication, and the incremental advances that derive from sharing ideas and discoveries, internationally, the stories that follow are written in no spirit of chauvinism or nationalism – although, as a scientist and a Scot, I have written them with a certain sense of pride.

My intention in the stories is to provide enough detail on the technical achievements of the subjects to satisfy a those with a professional interest. However, my intended audience is a general one, and I have deliberately avoided, for example, any complex mathematics or even detailed numbered references to the source material. I have taken care to describe some of the historical context in which the subjects lived and loved. My own Andersonian, St Andrean, humanist and Whiggish biases might be detected in some of the chapters, for which I make no apologies. One regret is that only a few women are specifically included. The obstacles experienced by women who wanted a scientific or medical career began only to be overcome in the later years of the 19th century, and even then with great difficulty, as exemplified by Elizabeth Garrett Anderson and the 'Edinburgh Seven'. But perhaps perceptive readers will notice how many of the male subjects of this book suffered from the early deaths of their fathers, only to benefit from the schooling and encouragement of enlightened mothers.

I hope that interested readers will also find a treasure-trove of reference material in the 'Sources and further reading' section, most of it available online, thanks to the scanning or digitisation efforts of the Internet Archive, Google, JSTOR, Hathi Trust and various university libraries. The research resources afforded by the World Wide Web are awe-inspiring and have been hugely helpful, especially in the last stages of writing the book, which was completed during the worldwide 'lockdown' response to the Covid-19 coronavirus pandemic of early 2020. As with previous epidemics and plagues, we will look to our scientists for lasting solutions, confident that science will continue to inform and enrich all our lives.

JM
Suffolk, England, May 2020

Part One: Out of the Dark Ages

1 Out of the Dark Ages

In the year 1600, Scotland's population was a meagre 800,000 but the country was well-endowed with seats of learning. Despite their locations on the north-west fringe of Europe, five Scottish universities were already well-established. The foundation of the University of St Andrews in 1410 had been prompted by the defection of Augustinian clergy from the University of Paris, as a result of the 'Avignon schism' in the Roman church, and from Oxford and Cambridge, as a result of the continuing hostility and sporadic warfare between Scotland and England. During the schism, Scotland supported the Avignon 'anti-popes' Clement VII and Benedict XIII. Bishop Henry Wardlaw of St Andrews had the full confidence of Benedict, who confirmed the establishment of the new university by Papal Bull in 1413. The University of Glasgow, founded in 1451, became the fourth-oldest university in the English-speaking world, followed by King's College, University of Aberdeen, in 1495. The Scottish Reformation led by George Wishart and John Knox resulted in the Edinburgh Parliament approving the Protestant confession of faith in 1560, rejecting any papal jurisdiction over the country's church or government. The 'rough wooing' of the young Mary Queen of Scots by English invasion, and her eventual forced abdication in 1567 after a tempestuous, tragic and divisive reign, firmly established Presbyterianism in Scotland. When the University of Edinburgh was founded in 1582, its authority derived not from Rome or Avignon, but from the Town Council and a Royal Charter granted by the Stuart King, James VI. In Aberdeen, the 5th Earl Marischal, George Keith, was ambitious to establish another 'Town College' based on the same principles. He founded the 'Marischal College and University of Aberdeen' in 1593, thus endowing that northernmost Scottish city with two universities. In 1603, with the accession of King James also to the English throne, Scotland and

England were united under a single monarch, but retained their separate parliaments, churches, laws and education systems.

The old Scottish universities admitted their students young, and based their education on a broad range of subjects: Latin, Greek, Law and Mathematics, as well as Arts, Divinity and Medicine (or *materia medica*, the very early forerunner of pharmacology). They looked mainly to the universities of continental Europe for their contacts, and it was common for Scottish students with sufficient means to supplement their studies in the likes of Paris, Leiden, Padua or Heidelberg. 'Sufficient means' of course usually required a relationship with a noble or landowning family. Natural philosophers - today we would call them physicists - mathematicians, astronomers and learned men of that ilk were often regarded with doubtful suspicion by the generally illiterate public, as probable necromancers, astrologers, alchemists and heretics. Natural philosophy could be a dangerous pastime. On February 17th, 1600, Galileo's contemporary, Giordano Bruno, was burned at the stake in Rome for heresy and maintaining the Copernican view that the Earth and planets orbit the Sun.

Through the 17th century, scholars in Scotland, as more widely in Europe, would be obliged to conduct their studies against a background of continual religious and political strife. The ruler of the newly-defined Great Britain, James VI and I, had no love for the Kirk's Presbyterianism and its obstreperous denial of the Divine Right of Kings. He snarled *"A Scottish Presbytery agreeth as well with a monarch as God and the Devil!"* When his son Charles I attempted to anglicise the Church of Scotland and reclaim ecclesiastical lands seized during the Protestant Reformation, the result was revolution. A 'National Covenant' pledging the defence of the Calvinist religion and the freedom of the Kirk was composed by Alexander Henderson, a minister at Leuchars in Fife. It was signed by 150 Scottish nobles in Edinburgh in February 1638, and became a national movement. Charles' attempts to suppress the Covenanters by force led directly to his downfall in the English Civil War - these days much more accurately called the Wars of the Three

Kingdoms - which engulfed all of Scotland, England and Ireland. The repercussions of the deposition and then restoration of the Stuart monarchy would continue to torment Scottish society for more than a hundred years.

2 The Logs and Bones of 'Marvellous Merchiston': John Napier (1550-1617)

Illustration 1: John Napier (unknown painter c. 1616)

Were logarithms discovered or invented? Either way the credit belongs to John Napier, a scholar born near Edinburgh in 1550. He was born to a noble family - as the 8th Laird of Merchiston he became a wealthy landowner and through his mathematical genius became known as 'Marvellous Merchiston'. As well as his achievements in mathematics, he was an astronomer and 'natural philosopher', now best known as the discoverer of logarithms, and the inventor of practical tools to exploit their use: logarithmic tables and the so-called 'Napier's bones'. He introduced and made common the use of the decimal point in arithmetic.

As was usual for members of the nobility at that time, he was privately tutored when young and did not have formal education until he was 13, when he was sent to St Salvator's College in the University of St Andrews. His stay in St Andrews was short - he dropped out of the University and travelled to mainland Europe to continue his studies, probably in France and Flanders.

In 1571, Napier, then aged 21, returned to Scotland. He married 16-year-old Elizabeth Stirling, daughter of James Stirling, the 4th Laird of Keir and of Cadder. The couple would have two children together. In his late twenties, he inherited his father's estates and he had a castle built at his estate at Gartness, Stirlingshire, in 1574. Elizabeth died five years later, and Napier later married Agnes Chisholm, with whom he had ten more children. At the castle of Gartness, Napier took an interest in the running of the land and explored his interests in the fields of religious politics, agriculture and mathematics. His father died in 1608, and Napier and his family moved to Merchiston Castle in Edinburgh, where he remained for the rest of his life.

Many mathematicians at the time were acutely aware of the problem of efficient computation. Napier focused some of his attention on inventing devices to address this problem and make computation faster and easier. In his memoirs Napier said *"Seeing there is nothing, that is so troublesome to mathematical practice, nor that doth more molest and hinder calculations, that the multiplications, divisions, square and cubical extractions of great numbers, which besides the tedious expense of time, are for the most part subject to many slippery errors, I began, therefore, to consider in my mind, by what certain and ready art I might remove these hindrances"*. He devised ingenious numbering rods which became known as 'Napier's Bones' that offered clever means to do just that. Napier's Bones are numbered rods which can be used to perform multiplication of any number by a number 2-9.

Napier coined a new term from two ancient Greek words: 'logos', meaning proportion, and 'arithmos', meaning number; compounding them to produce the word 'logarithm'. What are logarithms and how did Napier use them? In simple terms a logarithm is the power to which a number must be raised in order to get some other number. For example, the base 10 logarithm of 100 is 2, because 10 raised to the power of 2 is 100. Therefore the log of 100 is 2, the log of 1000 is 3, and so on. The power of logarithms is that many mathematical operations can be reduced to just the addition and subtraction of logarithms. Napier worked to calculate all the intermediate logarithms and tabulated them. As many generations of maths students can remember, log tables were an essential part of their equipment before electronic calculators arrived.

Illustration 2: Napier's 'Bones' (Science Museum, London)

For over 20 years Napier worked on developing methods to simplify calculations, and in 1614 he published *'Mirifici logarithmorum canonis descriptio'*, (a description of the wonderful table of logarithms), dedicated to Prince Charles, later King Charles I. It had a huge impact, enabling rapid calculations to be made by marine navigators, astronomers, land surveyors and book-keepers. A copy was sent to Henry Briggs, professor of mathematics at Gresham College in London. Briggs made Napier's method even easier by setting the log of 1 at zero. Napier agreed with the suggestion but, due to ill health, left the responsibility of setting up the new logarithm tables to Briggs. The new tables were published in 1624 and were called the Table of Common Logarithms. Napier was in contact with a network of European scholars, including Johannes Kepler and Tycho Brahe, and Kepler published his own work on logarithms in 1624.

Napier also saw the potential of other developments in mathematics, particularly those of decimal fractions, and symbolic index arithmetic, to tackle the issue of reducing computation. He knew that practitioners who had to make laborious computations generally did them in the context of trigonometry. Therefore, as well as developing the logarithmic relations, Napier set them in a trigonometric context so it would be even more relevant. But he was often perceived by the superstitious as a magician, a thaumaturge, a performer of miracles, who was believed to dabble in alchemy and necromancy. It was said that he would travel about with a black spider in a small box, and that his black rooster was his familiar spirit. In fact Napier was an ardent Protestant

Christian and in 1594 he had published *'A Plaine Discovery of the Whole Revelation of St. John'*, his interpretation of the Book of Revelations, and predicted that the end of the world would occur in either 1688 or 1700. His anti-papal stance also fuelled his concern that Philip of Spain might invade Scotland. And just in case there was a war with Catholic Spain, Napier invented four new kinds of weapons: an artillery piece; a type of battle tank, covered with plates of metal and driven by men inside; and two kinds of mirrors for setting fire to enemy ships.

Napier's ingenuity commanded widespread amazement and admiration, and perhaps the fame of his reputed magical powers took him into unfamiliar jobs. It is said that a contract still exists made between Napier and Robert Logan of Restalrig, a notorious Scottish knight, for a treasure hunt. Napier was to search Logan's home in Berwickshire, Fast Castle, for treasure allegedly hidden there, wherein it is stated that Napier should *"do his utmost diligence to search and seek out, and by all craft and ingyne to find out the same, or make it sure that no such thing has been there."* It seems the business relationship went badly. The terms of a lease granted by Napier at his Gartness estate, in September 1596, expressly stipulated that the lessee should neither directly nor indirectly suffer or permit any person bearing the name of Logan to enter into possession.

Napier died at home in Merchiston Castle from the effects of gout at the age of 67. His birthplace, Merchiston Tower, is now part of Edinburgh Napier University. His remains were buried at St Giles Cathedral, though after the loss of the kirkyard there to build the Scottish Parliament House in 1641, he was memorialised at St Cuthbert's Church in the west side of Edinburgh. His name is given to the title of a lunar crater, and to the alternative unit to the decibel, the Neper, used particularly in electrical engineering.

3 Scientist, freemason, soldier, spy - and founder of the Royal Society: Robert Moray (1609-1673)

The future Sir Robert Moray was born to the noble family of Sir Mungo Moray in Perthshire, in 1608 or 1609, and had a truly astonishing life as a soldier of fortune, freemason, spy, diplomat, statesman... and ultimately as the key founder of the Royal Society. He was active in the Covenanter rebellion against the attempt of Charles I to impose Anglicanism on the Scottish church, yet later became a confidante of

Illustration 3: Masonic mark of Sir Robert Moray

both Charles I and his son Charles II. Robert's brother, Sir William Moray was to become Master of Works for Charles II. In his most significant and lasting contribution to science, Robert convened the meeting of the 1660 'committee of twelve' on 28th November 1660 and arranged the Royal Charter that led to the formation of the Royal Society, the first scientific body of its kind. How and why this happened is a fascinating story of political turmoil and intrigue.

Moray took a keen interest in 'natural philosophy' and 'mechanics' from early in life. In 1623 he visited the artificial island constructed to mine coal from under the Firth of Forth at Culross by Sir George Bruce (grandfather of Alexander Bruce, who became Moray's close friend). In a letter written later in his life, Moray cites 1627 as the year when he began to study *"to understand and regulate my passions"*. As a boy, Moray was captivated by civil engineering partly inspired by the visit to the undersea mine. After completing his education he became a soldier and then a diplomat and politician. Some sources claim that Moray attended the University of St Andrews (though his name does not appear in matriculation records) and that he continued his university education in France. However Moray himself wrote to Alexander Bruce (who definitely did attend St Andrews University), jokingly proposing a debate, in which Moray said he would force Bruce to *"rub up your St Andrews language"*, and threatening that *"one may give you your hands full that was scarcely ever farrer East then Cowper"* (the town of Cupar which lies a few miles to the west of St Andrews). Our main interest in Moray's life is in his contribution to science, but his life was so adventurous that it begs to be described more fully.

At the age of 24 Moray enlisted as a soldier of fortune and officer in the army of Louis XIII, probably in the Scottish regiment formed by Sir John Hepburn in 1633. Moray's talents - linguistic, military and diplomatic - were quickly recognised by Louis' chief minister Cardinal Richelieu. By 1638 in Scotland, fierce resistance had grown against the attempts of Charles I to impose Anglican forms of worship on the Presbyterian Scots, and the National Covenant was published pledging allegiance both to the Presbyterian faith and the King - a dangerous and ultimately unsustainable balance. In France, Moray had become a favourite of Richelieu and was groomed as a spy to join and inform the Cardinal on the Scottish Covenanter rebellion. Richelieu's motivation seems twofold: first to gain intelligence on the military, religious and political situation in England and Scotland; and secondly to ensure that events in Great Britain did not unfold to the disadvantage of France, which was engaged in a 30 Year War against Spain, the Habsburgs, and the Holy Roman Empire. In any event, Cardinal Richelieu chose Moray to promote the interests of France, and in 1638 sent him back to Scotland, ostensibly to recruit more soldiers of fortune for the French army. Moray had by then become persuaded by the anti-Anglican movement in Scotland and he joined the Covenanters army in

Edinburgh where his military experience and engineering ability led to his appointment as quartermaster-general in the Covenanting army that invaded England in 1640 in the Second Bishops' War against Charles I. Moray had responsibility for fortifications and the army camp constructions, and seems to have been effective. The Scottish Covenanting army overran Northumbria, County Durham and Newcastle-upon-Tyne in August 1640, defeating Charles' army in the Battle of Newburn.

It was while garrisoned in Newcastle in May 1641 that Moray was initiated into Freemasonry, the fraternal society that had its strong roots in 16th century Scotland. The Masonic officers who initiated him were General Alexander Hamilton (commander of the Covenanters' Army in Newcastle) and John Mylne, Master Mason to King Charles I. Although Moray was initiated into the Scottish lodge, this was the first initiation into Freemasonry on English soil. Thereafter, Moray regularly used a five-pointed star, his masonic mark, on his correspondence. Freemasonry encouraged his innate love of symbolism and helped him to develop distinct and individualistic ideas throughout his life, and he earned a reputation for original thinking. He once wrote of himself: *"I have been reported to be writing against Scripture, an Atheist, a Magician or Necromancer, and a malignant - for ought I know by half a Kingdom."* Years later, Charles II was to remark teasingly that he believed Moray to be head of his own church.

Moray's skills in diplomacy, mediation and self-advancement became even more evident at this time. In 1642 he was appointed Lieutenant-Colonel in the newly re-formed regiment of Scottish Guards in the French army - the *Garde Ecossaise* - while also acting as a liaison officer between the Scottish Covenanters' Army and the court of Charles I at Oxford. Charles must have been impressed, and he was keen to achieve a settlement with the Covenanters. Moray was knighted by him on 10th January 1643.

Later that year, the newly-knighted Sir Robert returned to France to take up his position in the *Garde Ecossaise,* who were already involved in battle in the 30 Years War, which France had entered on the side of the Protestant states arranged against the encroaching Habsburg empire of Spain and Austria. The Scottish Guards reinforced the French army and crossed the Rhine on 1st November, capturing the strategic town of Rottweil in southwest Germany. The march towards Tuttlingen on the Danube was much less successful: the French army was completely defeated by the Habsburg armies and several of the *Gardes Ecossaise* - including Moray - were captured and imprisoned in Ingolstadt in Bavaria. France did not agree to an exchange of prisoners, and there was no hope of immediate freedom. Much of Moray's imprisonment was spent in study - including the study of magnetism as described in a book by the German Jesuit scholar Kircherus, with whom Moray corresponded from jail.

After 17 months in prison, Moray was unexpectedly released at the end of April, 1645. The ransom of 16,500 Scots pounds was paid by Robert Murray, at that time a merchant in Paris and later Sir Robert Murray of Cameron and Priestfield, and Lord Provost of Edinburgh. Our Sir Robert Moray was never again on active service in France, though on the death of James Campbell, Earl Irvine, Moray took full command of the *Garde Ecossaise*. Meanwhile, the political situation in Great Britain had evolved dangerously.

When Moray had left for France in the Spring of 1643, Civil War had already erupted in England between Royalists loyal to Charles I and the forces of the English Parliament which had had enough of Charles, his Royal Prerogatives and autocratic rule. A 'Solemn League and Covenant' had been drafted in which the Scots agreed to support the English Parliamentarians in their dispute with the Royalists, in return for a pledge for a civil and religious union of England, Scotland, and Ireland along Presbyterian and Parliamentary lines. It was accepted by the Scottish Parliament and Church of Scotland in August and by the English Parliament in September of 1643. As a result the Scottish army again invaded England in January 1644 in support of the English

Parliament, and engaged in sustained warfare that resulted in the capture of Charles I at Nasby in June of 1645.

The King's defeat and capture required a shift in the policy of the French. Mazarin, the chief minister who had succeeded Richelieu, had no desire to see a republic established in England. He wanted the British monarchy to be preserved, but in accommodation with the English Parliament and the Scottish Presbyterians. Again the diplomatic skill of the newly-released Sir Robert Moray was enlisted. It seems likely that his release from prison in Bavaria was arranged to permit him to engage in just such mediation between Charles I, the English Parliamentarians, and the Scottish Covenanters. For the next two years Moray was engaged in complex shuttle diplomacy between London, Paris and Edinburgh, trying to reconcile the parties. He persuaded Charles to put himself in the hands of the Scots army in Newcastle, since he and Charles believed the Covenanter commitment to Presbyterianism could be mollified. It could not. The Scots insisted that Charles personally signed the Covenant, and he refused. After tortuous negotiations, there was impasse, and Charles had exhausted the patience of every party concerned. In December 1646, the Scots army agreed to hand Charles back to the custody of the English Parliament. Charles decided to escape to the Continent - dressed as a serving-woman - in a ship that was moored in Tynemouth. In the words of one source: *"... Sir Robert Moray was to have conveyed the King thither in a disguise, and it proceeded so far that the King put himself in the disguise and went down the back stairs with Sir R. Moray. But His Majesty, apprehending that it was scarce possible to pass through the guards without being discovered, and judging it hugely indecent to be caught in such a condition, changed his resolution and went back..."*

The Scots army handed Charles over to the English and returned to Scotland in January 1647, in return for settlement of their due expenses in the English campaign. A faction of the Covenanters, called the Engagers, continued to support the Royalist cause, and Moray was among them. Most of Moray's effort at this time was further recruitment of soldiers for the *Garde Ecossaise*, while liaising with the

Prince of Wales, who was exiled in France, to persuade the Prince to return to Scotland.

After his detention by the English Parliament, the King was approached by the Scots, the Presbyterians in the English Parliament, and the Grandees of the New Model Army, all attempting to reach an accommodation with him that would achieve peace and preserve the Crown. But then, the breach between the New Model Army and Parliament widened, until the Presbyterians in Parliament, with allies among the Scots and the remaining Royalists, saw themselves strong enough to challenge the Army, which began the Second English Civil War. Subsequently, the Grandees and their civilian supporters would not reconcile with the king or the Presbyterian majority in England's Parliament. The Grandees acted; soldiers were used to purge the English Parliament of those who opposed the Army. The resulting 'Rump Parliament' of the Long Parliament then passed enabling legislation for putting Charles I on trial for treason. He was found guilty of treason against the English commons and was executed by beheading on the 30th of January 1649.

The Scottish Parliament proclaimed the Prince of Wales as King Charles II of Scotland in February 1649, just six days after his father's execution. England entered the period known as the English Commonwealth, and became a *de facto* republic led by Oliver Cromwell. Cromwell defeated the Royalist army of Charles II (of which the majority were Scottish Engagers) at the Battle of Worcester in September 1651, and Charles, accused of treason against England, fled to mainland Europe. Cromwell became a virtual dictator of England, Scotland and Ireland. Charles spent the next nine years in exile in France, the Dutch Republic and the Spanish Netherlands.

During this turbulent period, Sir Robert Moray was a more or less active Royalist, advising the new King, and supporting the Royalist movements in Scotland. He had married Sophie Lindsay, daughter of Lord Balcarres, but in the January of 1653, Lady Moray died in childbirth, and the child was stillborn. Moray was said to have sat constantly at his wife's bedside, feeling her pulse, and encouraging her to withstand the pain; *"he did speak so excellently to her as did exceed by far what the best ministers said who frequently came to her."*

Moray appears to have moved on quickly from this double tragedy. In 1653 and 1654, he was partly involved in an unsuccessful military revolt against Cromwell known as the Glencairn Rising, yet he was also accused of treason against Charles II, and was briefly imprisoned in Scotland. The 'evidence' of his supposed treasonous plotting was based on letters shown later to have been forged by his enemies. On his release and given the defeat of the Glencairn Rising, Moray returned to the Continent in 1654 and for the next five years lived in Bruges, The Hague, Maastricht, and finally Paris, joining Charles II in his own exile. In England however, the political crisis that followed the death of Cromwell in 1658 resulted in the restoration of the monarchy. Charles was invited to return, and on 29th May 1660, Charles' 30th birthday, he was received in London to public acclaim. After 1660, all legal documents were dated as if he had succeeded his father as King in 1649.

Much of Moray's time in Continental exile was spent in study and in scientific explorations (he wrote *"you never saw such a shop as my laboratory is"*) including the phenomena of magnetism and the laws of motion - though his wider fields of study are unclear. On his return to Britain he was appointed to several political positions in the Scottish government - and became a Lord of the Exchequer, a Privy Councillor, Deputy Secretary for Scotland, and a Lord Ordinary of the Court of Session - though he did not return to reside in Scotland until 1667. Much of his focus was instead on the discussions of scientists and thinkers who met regularly in Oxford and at Gresham College in London. These included Robert Boyle, Christopher Wren, Alexander Bruce and William Brouncker. On the 28th November 1660, at Gresham College, Moray was one of those present at a notable meeting at which Wren, the Gresham Professor of Astronomy, delivered a lecture. The twelve men in attendance were a carefully chosen mix of four Royalists (William Brouncker, 2nd Viscount Brouncker; Alexander Bruce, 2nd Earl of

Kincardine; Sir Paul Neile; and William Balle), six Parliamentarians (John Wilkins, Robert Boyle, Jonathan Goddard, William Petty, Lawrence Rook, Christopher Wren) and two others with more flexible views, Abraham Hill and Moray. The meeting resolved to establish a more permanent society for the *"promoting of Physico-Mathematical Learning"* and agreed that Royal endorsement of the proposal was highly desirable.

Moray's close relationship with Charles II was exploited and within a week Moray reported that the King approved the proposal. He briefly assumed Presidency of the still informal society before passing the presidency to William, Lord Brouncker, while Moray pursued the King for the ratification of the Royal Charter and Seals which confirmed the establishment of the Royal Society. That was achieved in August 1662, and in subsequent revisions of the Charter as it is today. Moray led in formulating the Society's rules and statutes, and in that month the Society resolved that *"Sir Robert Moray should be thanked for his concern and care in promoting the constitution of the Society into a Corporation"*. The Royal Society still holds its Annual General Meeting on Saint Andrew's Day (30th November) in apparent acknowledgement of Moray's importance in the formation of the society.

Moray's concern to achieve Royal endorsement for the Society was mainly to gain the financing necessary to achieve its objectives. The Society did not have enough money to conduct serious experiments or in-depth studies, and urgent engineering problems had to be addressed. When the First Charter was achieved, Moray stepped back from the Presidency, and astutely promoted the Naval enthusiast, Lord Bouncker, as the First President, intending that the Society would continue under its own steam with Royal funding to focus on solving the technical and ship design problems then plaguing the Royal Navy. In October 1662, the King ordered the Lord Lieutenant of Ireland to *"make a grant accordingly to R. Boyle and Sir R Moray...in trust for the Royal Society"*. The order was never obeyed, and the King did not force the matter. In February 1664 Sir Robert moved in the Society's Council *"that every one of the Council might think on ways to raise a revenue for carrying on the design and work of the Society"*. Nevertheless shortage of funds was to be a problem for the Society for many years.

Moray continued to serve the Royal Society in both a managerial and scientific capacity, though he moved away from London to avoid the Plague and returned to Scotland to fulfil his government duties there. He was not satisfied with the conduct of experiments in the Society, which was haphazard, and helped to form two committees to oversee a more structured approach. His own experiments and observations were interesting but not of critical importance, covering such subjects as a peculiar water spring near Chertsey in England; the anomalous tides in the Outer Hebrides; and wood-worms in ash timber. He was commissioned by the Society to write a History of Masonry, which, if it was ever completed, does not exist in the Society's records. He continued his correspondence on many subjects, including the progress of the Royal Society, with many scientists including Christiaan Huygens. In the 1660s Huygens was arguably the most able scientist in Europe, and knowledge in Britain of Huygen's work was chiefly gained through his correspondence with Moray. It worked both ways. Huygens valued the scientific news from Britain contained in Moray's communications, and Moray's comments on his own work. Huygens spoke of Moray as the 'soul' of the Royal Society.

In later life, as Moray's activity declined, he wrote fewer letters and engaged less in the Society's proceedings. He became more reclusive, and by the time of his death, aged 64, his finances were nearly exhausted. His death was almost as dramatic as his life. Following a visit from John Aubrey, the antiquarian and natural philosopher, and dinner with Lord Ashley, the 2nd Earl of Shaftesbury, Moray was taken suddenly ill at his home in London, possibly from food poisoning, and died the same evening.

No definitive images of Moray exist, and his portrait is sometimes confused with that of his patron, King Charles II. Moray was buried at the order of King Charles, and at the King's expense, in the Poet's Corner of Westminster Abbey.

4 The maritime pendulum clock: Alexander Bruce (1629-1681)

Alexander Bruce was a contemporary and close friend of Sir Robert Moray, and as the 2nd Earl of Kincardine, was often known simply as 'Kincardine'. Like Moray, he was a politician, judge and freemason, and a member of the 'Committee of Twelve' who in 1660 formed what soon became the Royal Society, in which he became a Fellow.

Alexander's grandfather, Sir George Bruce had built a fortune from coal-mining and salt-production, using an artificial island

Illustration 4: Alexander Bruce, 2nd Earl Kincardine (by Johannes Mijtens)

constructed in the Firth of Forth near Culross, and built Culross Palace in Fife, in 1597. Alexander's elder brother, Edward, was made 1st Earl of Kincardine by Charles I in 1647. As a supporter of the Royalist cause during the English Civil War and the Wars of the Three Kingdoms, Alexander was forced to flee and took refuge in the Low Countries. There he joined the court of Elizabeth Stuart, eldest daughter of King James VI and the so-called 'Winter Queen' due to the brief reign of her husband, Frederick V, as King of Bohemia. While in exile, Alexander, aged 30, married Veronica van Sommelsdijck, a sister of a prominent Dutch politician and aristocrat. When his brother Edward died unmarried in 1662, he succeeded to the Earl's title and the Kincardine estates.

Bruce became extremely interested in the problem of longitude - that is, the practical problem of accurate time-keeping at sea which was essential to the measurement of the longitudinal position of ships. To determine longitude at sea, a comparison is needed between the time aboard ship, and also the time at the home port (or another place of known longitude) at the same moment. The difference in time allows the calculation of east-west separation between the two points. Since the earth takes 24 hours to complete a revolution of 360 degrees, a difference of one hour corresponds to 15 degrees of longitude. The corresponding distance depends on the ship's latitude, which is more easily determined by measuring the length of the day, the peak height of the Sun, or the height of known stars above the horizon. At the equator, 15 degrees of longitude corresponds to around 1,600 kilometres, but north or south of that, each degree of longitude has a lower value of distance shrinking from 109 km at the equator to zero at the poles. Precise knowledge of the time aboard ship was difficult to achieve. In the era of pendulum clocks, in a rolling and heaving ship, clocks would run fast or slow, or stop altogether. The epic voyages of Vasco de Gama, Magellan and Drake had only been achieved through a combination of seamanship, dead reckoning, and good luck.

For a solution, large rewards were on offer from most of the European monarchs. The Dutch astronomer, mathematician and inventor Christiaan Huygens had devoted much thought to the problem, describing in detail the physics of pendulums, and arrived at the idea of the pendulum clock building on the insights of Galileo. His first pendulum-regulated clock was built in 1656, but the problem for making a marine-capable clock remained.

The first practical marine pendulum clock emerged from the collaboration of Huygens and Alexander Bruce during November and December 1662. Huygens had experimented with various ways of suspending clocks from pivots to compensate for the ship's motion.

Bruce had the idea of encasing a steel ball in a brass cylinder to achieve the same compensation and commissioned two clocks from the Dutch instrument maker Severyn Oosterwijck, which proved to work well in tests at The Hague. The clocks were transported by Bruce to London in 1663. The support and influence of the Royal Society led to sea trials conducted by Captain Robert Holmes on voyages first to Lisbon, then to Guinea in Africa and onward out into the Atlantic. The first voyage produced encouraging data, including clocked longitudes that agreed well with accepted values in well-known waters. The second voyage in 1664 proved to be more dramatic.

Illustration 5: The Bruce-Oosterwijck Longitude Pendulum Clock (National Museums Scotland)

Captain Holmes set sail west from the island of St. Thomas off the African coast, and travelled 800 leagues (more than 4,000 kilometres) before taking a course northeast back toward the African coast. After nearly a week, fresh water supplies ran low. Holmes' fellow masters urged that he should head for the Barbados Islands. They reckoned that the squadron was still some 100 leagues from the Cape Verde Islands in mid-Atlantic. However Holmes's clock-time combined with a timing of the local noon placed him only 30 leagues (166 km) away, just a day's run. Holmes continued on course and found land during the next afternoon. Bruce, Moray, the Royal Society and later Huygens learned that the ships had been becalmed for a while after heading northeast and had drifted with currents around 80 leagues eastward, motion that would not have been detected by traditional dead-reckoning navigation.

Huygens published an account of Holmes's successful trials in February 1665 and put the marine clocks on public sale along with a 'Brief Instruction', on their operation and use in determining longitude.

Bruce claimed a part in the marine clocks' design and a share in any profits from it. He later attempted to file for an English patent. Huygens avoided any explicit discussion of the clock design and of Bruce's contribution to it. Only in Huygens' *'Horologium oscillatorium'* is there an overview of the first marine clock design: *"Instead of a weight they had a steel strip wound in a spiral, but the force of which the wheels were turned 'round, just as is commonly employed in those small watches that are wont to be carried about. So that the clocks could endure the tossing of the ship, he [Bruce] suspended them from a steel ball enclosed in a brass cylinder, and extending downward the arm of the crutch that sustains the pendulum's motion he doubled it to resemble the form of an inverted letter F; namely, lest the pendulum's motion wander out in a circle with the danger of stoppage."* Huygens did not want to divulge any more details of the clock, and he hoped soon to have more data to prove further the seaworthiness of the design.

In fact, further sea trials ordered by the French Academie des Sciences proved disastrous. Although tests in 1668 and 69 successfully measured the longitude difference between Crete and Toulon in France, longer distance voyages to the West Indies in 1670 proved fatal to the project. Members of the Royal Society (perhaps even including Bruce) had begun to doubt that the clocks could remain truly stable in a rolling ship, and shortly after the beginning of the 1670 sea trials, one clock stopped in a storm, and a second stopped soon after. The trials were abandoned. Although Huygens continued to develop various ingenious clock designs, it was not until the inventions of the English horologist John Harrison, 70 years later, that truly accurate seaworthy clocks were developed.

Aside from his work on clocks, Bruce's scientific and technical interests were mainly in commercial developments that might advance his businesses. Much of his correspondence with Sir Robert Moray concerned very practical subjects such as the efficient production of coke, bricks, fish-breeding, and the practices of European water- and windmills. However Bruce's involvement in Scottish Royalist and Covenanter politics was to the detriment of his business interests. By

1667 he was a Privy Councillor to the King, a Joint Commissioner of the Treasury of Scotland, and an Extraordinary Lord of Session (that is, a lay member of the judicial Court of Session). But his own businesses declined, and by his death in 1680 he was much impoverished. His friend and relative, Bishop Gilbert Burnet, wrote of him (in Burnet's 'History of my Own Times'):

" He was both the wisest and worthiest man of his own country, and fit for governing any affairs but his own...and he was very capable of it, having gone far in mathematics, and being a master of mechanics...a deep judgment appeared in everything he said or did... he was a faithful friend and a merciful enemy."

5 Royal botanist and physician: Robert Morison (1620-1683)

Illustration 6: Woodcut of Robert Morison

Born in Aberdeen, Robert Morison was an outstanding scholar and gained his Master of Arts degree from Marischal College at the age of 18. He supported the Royalist cause in the Wars of the Three Kingdoms, and at the age of 19 was wounded in the head at the Battle of the Bridge of Dee between Royalists led by Viscount Aboyne and the Covenanters of James Graham, Marquess of Montrose. Morison recovered and fled to France when it became apparent that the Royalist cause in Scotland was lost to the strengthening Covenanter movement.

In 1648 Morison was awarded a doctorate in medicine at the University of Angers in France, and studied in Paris under Vespasien Robin, botanist to King Louis XIV. Morison immersed himself in the study of botany, and on the recommendation of Robin he became director of the French Royal Gardens at Blois. He held this post for 10 years and travelled extensively within France, searching for new garden plant species and refining his ideas on plant classification. His royal connections were established when he became physician and botanist to Louis XIV's brother, Gaston, Duc d'Orleans. Following the Restoration of Charles II to the English throne In 1660, Morison moved to England and became physician to the King, as well as acting as his botanist and superintendent to all the royal gardens, rewarded with a grace-and-favour house and an annual salary of £200.

Decades earlier, the University of Oxford had been bequeathed £250 for the purchase of land for a 'Physic Garden' (devoted to medicinal plants) and also 'certain revenues' to fund a Chair of Botany at Oxford. In 1669 Robert Morison became Oxford's first Professor of Botany, a post that he held for 14 years. One of his first publications for the Oxford University press was the *'Hortus Regius Blesensis'*, a catalogue of the Blois garden to which Morison added the description of 260 plants that were previously largely unidentified. In it Morison referred to the antiquity and perfection of plants, which he believed had survived in their original forms as made by God on the third day of the creation. He gave many of his first courses of lectures in the physic garden, at a table covered in plant specimens, three times a week for five weeks, attracting sizable audiences and becoming part of the intellectual tours of foreign dignitaries. On the 1683 visit of James, Duke of York, the University took the opportunity to show its academic accomplishments and its political and religious loyalties to the Catholic heir to the throne. Morison's accent seemed to cause problems. His Oxford contemporary, the antiquary Anthony Wood, remarked that on that occasion Morison *"speaking an English speech also, was often out and made them laugh. This person, though a master in speaking and writing the Latin tongue, yet hath no command of the English, as being much spoyled by his Scottish tone"*.

In his work at Oxford, Morison clarified and developed the first systematic classification of plants, building on the thoughts of Andrea Cesalpino's *'De Plantis'* and laying the foundation of the later classification system of Carl Linneus, 50 years later. Morison drew criticism from his contemporaries as he stated that he had derived his plant classification schema from "the book of Nature alone" and did not make reference to Cesalpino, whose ideas he built on. Although he did not cite Cesalpino's work, a copy of *'De Plantis'*, heavily annotated by Morison, is in the Oxford Botanic gardens. Hence, speaking of Morison, Linnaeus was equivocal in his praise. They could never have met, but Linnaeus wrote of his predecessor: *"Morison was vain, yet he cannot be sufficiently praised for having revived [a] system which was half expiring. If you look through Tournefort's genera you will readily admit how much he owes to Morison, full as much as the latter was indebted to Cesalpino ... All that is good in Morison is taken from Cesalpino, from whose guidance he wanders....'*

Morison's main focus at Oxford was the production of his *magnum opus* herbal reference, the *'Plantarum Historiae Universalis Oxoniensis'*. It proved to be a laborious and costly undertaking. Morison impoverished himself in the preparation just of the first and only volume that appeared in his lifetime. Friends and donors provided the cost of its 126 plates of illustrations, and he borrowed three loans of £200 each from the University press. The work was to have been in three parts: the first part was to be devoted to Trees and Shrubs, and the other two parts to the Herbs. The volume published in 1680 deals with only five out of the fifteen sections into which he classified herbaceous plants. In the preface he gives as his reason for beginning with the Herbs rather than with the Trees and Shrubs, that he wished to accomplish first the most difficult part of his task lest, in the event of his death before the completion of the *Historiae*, it should fall into the hands of incompetent persons. He did not live to finish his great undertaking. In November, 1683, he was seriously injured by the pole of a horse-drawn carriage as he was crossing the Strand near Charing Cross in London, and died the following day, aged 43. He was buried in the church of St Martin-in-the-Fields.

After his death, his papers passed to Jacob Bobart, the keeper of the Oxford Physic Garden. They included a further four sections of descriptions of herbs for the *Plantarum*. Bobart added his own descriptions and had them translated into Latin to complete the 15 sections that Morison had planned. *'Plantarum Historiae Universalis*

Oxoniensis' was eventually published by the Oxford University press in 1699. It did not sell well, and the University's loss was reckoned to be more than £2000. Morison's widow, described as "a sharp Gentlewoman", was unwilling, or more likely unable, to repay the debts to the University press. Her representatives agreed to a settlement.

Morison's plant taxonomy was largely superseded by the work of later botanists, and the limited adoption of his work might have been due to his relative isolation and reluctance to collaborate. Yet Carl Linnaeus regarded Morison as the principal pioneer of taxonomy in botany. The physician Robert Gray, a Jacobite Royalist supporter who helped to produce a biography of Morison by Alexander Pitcairne in 1697, wrote of him: *"he was communicative of his knowledge, a true friend, an honest countreyman, true to his religion whom neither the faire promises of the papists nor the threatenings of others could prevail upon to alter or change".*

6 Telescopes and trigonometry: James Gregory (1638-1675)

James Gregory was the youngest of three children of John Gregory, an Episcopalian Church of Scotland minister, who was persecuted for opposing the Covenanters. James was born in the manse at Drumoak, a few miles from Aberdeen. He came from an amazing family already noted for their ability in mathematics and practical engineering. His grandfather, David Anderson, was known as 'Davie do a' thing', and the 'Archimedes of Aberdeen'. Davie reputedly constructed the spire of the Kirk of St Nicholas, (Aberdeen's 'Mither Kirk') installing with his own hands its weather-cock; and he had removed a dangerous submerged rock from the mouth of Aberdeen harbour by harnessing it to the tide. It was Davie's daughter, Janet, who had married John Gregory (sometimes spelled Gregorie) of Aberdeen, who had himself studied in the University.

Illustration 7: James Gregory - portrait by John Scougal (National Trust for Scotland, Fyvie Castle)

James Gregory received his first lessons in geometry from his mother Janet. His brother, David, some ten years his senior, directed his education after his father's death in 1651, and gave him a copy of

Euclid's *'Elements'*. James attended the Aberdeen Grammar School and then graduated from Marischal College, University of Aberdeen. At that time he suffered sickness from 'the Quartan Ague' - a form of malaria where fever recurs every three or four days, and years later wrote: *"It is a disease I am happily acquainted with, for since that time I never had the least indisposition; nevertheless ... I was of a very tender and sickly constitution formerly."*

His first book, the *'Optica Promota' ('The Advance of Optics')* brought him fame at the age of 24. It described the geometry of mirrors and lenses, and contained a design of the earliest reflecting telescope. Reflecting telescopes consist of a combination of mirrors and lenses, which greatly reduce the length of the tube, by a factor of 4 or more, compared to a refractive telescope which consists of lenses alone. In 1663 Gregory went to London, to have the book published, and met Robert Moray, and probably other founders of the Royal Society. A large objective mirror was commissioned to be made in London to Gregory's specification. But the manufacturing methods of the time were incapable of the precision needed, and the project was abandoned. It fell to Isaac Newton, 9 years later, to demonstrate to the Royal Society a working reflecting telescope, only 6 inches long.

In 1664 Gregory departed for the University of Padua, travelling on the way through European centres of learning in Flanders and Rome and returning via Paris. In Padua Gregory worked with expert Italian geometers including Bonaventura Cavalieri, who made developments in geometry that were precursors to integral calculus. At Padua, Gregory wrote two brilliant works, *'Vera Quadratura' - ('The true quadrature of the circle and hyperbola')* - and the *'Geometriae Pars Universalis' ("The Universal Part of Geometry")*, which reinforced his reputation as a great mathematician. The *'Vera Quadratura'* attempted to prove various hypotheses on the area of conic sections and elucidated the properties of the 'transcendental' number *'e'*. It attracted the hostility (and perhaps the jealousy) of the great Flemish physicist, Huygens, who accused Gregory of using his own results without acknowledgment - some of which Gregory had never even seen.

On Gregory's return to London in 1668 he was awarded the Fellowship of the Royal Society, and presented to them several papers on mechanics and gravitation. Soon, probably due to the persuasion of Sir Robert Moray, King Charles II endowed a new Regius Chair of Mathematics at the University of St Andrews, which was designed to give Gregory the scope to continue to express his genius. He took up this post in the autumn of 1668, just 30 years old, a mathematical whizz-kid bursting into the cloistered calm of that ancient university, where the advances of Huygens, Kepler, Descartes and Galileo were as yet serenely ignored. Around that time Gregory wrote *"I am now much taken up, and have been so all this winter by past, both with my public lectures, which I have twice a week, and resolving doubts which some gentlemen and scholars propose to me. This I must comply with, nevertheless I am often troubled with great impertinences: all persons here being ignorant of these things..."*

More happily, Gregory was married in 1669 to Mary Jamesone, a young widow and daughter of an artist. It was said that Mary inherited her mother's beauty and her father's artistic talents, and the couple had two daughters and a son, also James, who became Professor of Physic (Medicine) at King's College, University of Aberdeen.

At St Andrews, Gregory *Senior* lectured, made astronomical observations, and constructed plans for the construction of an Observatory. Gregory and Huygens were once more good friends, and Huygens had recommended Gregory to Louis XIV of France for a pension and an appointment in Paris in recognition of his talents. Gregory though remained in Scotland and continued to make acute observations, and was the first to record the effects of light diffraction through a grating by using a sea-bird feather to view sunlight. He wrote *"Let in the sun's light, by a small hole to a darkened house, and at the hole place a feather (the more delicate and white the better for this purpose), and it shall direct to a white wall or paper opposite to it a number of small circles and ovals (if I mistake them not) whereof one is somewhat white (to wit, the middle which is opposite the sun) and all the rest severally coloured. I would gladly hear Mr Newton's thoughts of it."* This was a year after Newton had observed diffraction using a prism, and the phenomenon was still highly controversial.

Around this time Gregory made important discoveries in differential calculus, and indeed seems to have been a co-discoverer of the calculus alongside Newton, Barrow and Leibniz. But Gregory's work was not published; and, on learning that Newton had anticipated him - and assuming modestly that Newton had progressed at least as much as himself (which he had not) Gregory decided to withhold his work until his young rival had published his own - which did not in fact take place until many years after Gregory's death.

The observatory at St Andrews was eventually built - the first of its kind in Scotland or England - but Gregory was now discontented at St Andrews, and was tempted away to the Chair of Mathematics at the University of Edinburgh: *"I was ashamed to answer, the affairs of the Observatory of St Andrews were in such a bad condition; the reason of which was, a prejudice the masters of the University did take at the mathematics, because some of their scholars, finding their courses and dictats opposed by what they had studied in the mathematics, did mock at their masters, and deride some of them publicly. After this, the servants of the colleges got orders not to wait on me at my observations: my salary was also kept back from me; and scholars of most eminent rank were violently kept from me, contrary to their own and their parents' wills, the masters persuading them that their brains were not able to endure it. These, and many other discouragements, obliged me to accept a call here to the College of Edinburgh, where my salary is nearly double, and my encouragements otherwise much greater".*

The move to Edinburgh University proved to be tragically short-lived. In 1675, around a year after assuming his new professorship, Gregory was blinded by a stroke while viewing the moons of Jupiter with his students. He died a few days later at the age of 36. The crater *'Gregory'* on the Moon is named after him, and the Gregorian telescope design remains the basis of many modern optical and radio telescopes. His reluctance to publish much of his brilliant work in mathematics means that he is relatively uncelebrated, given that he derived many of the results of Maclaurin, Taylor and Newton years before them. And it was said *"For all his talent and promise of future achievement, Gregory did not live long enough to make the major discovery which would have gained him popular fame. For his reluctance to publish his 'several universal methods in geometry and analysis' when he heard...of Newton's own advances in calculus and infinite series, he posthumously paid a heavy price ..."*. His relatives and descendants however would hold professorial Chairs in Scottish universities in an almost unbroken sequence for two centuries. His nephew David held the chair of mathematics at Edinburgh from 1683-91. James' son James junior, who was professor of physic (medicine) at King's College in the University of Aberdeen, had in turn two sons, John and James, who were also distinguished medical men. John (1724-1773) would be first physician in Scotland to George III, while James (1753-1821) became professor of medicine at Aberdeen and then Edinburgh, and was one of the founders of the Royal Society of Edinburgh. He had five daughters and six sons, including James Crawford, a doctor; Donald, a historian and antiquary; Duncan, a mathematician; and William (1803-1858), who was professor of chemistry at the Andersonian University in Glasgow, and then at both Aberdeen and Edinburgh. The Gregorian dynasty would become as famous as the Gregorian telescope.

Part Two: The Scottish Enlightenment

7 The Scottish Enlightenment

The Scottish Enlightenment sprang from very unpromising beginnings, and flourished despite the political turmoil of much of the 18th century. Scotland was already ravaged by the civil wars of the middle 1600s. The last years of the 17th century, and the first of the 18th, continued to be hard on the country. The religious extremism of the Covenanters was a constant source of deadly warfare with the supporters of Charles II - the 1680s became known as 'the Killing Time'. After the death of Charles in 1685 and the short two-year reign of his Catholic brother, James VII of Scotland and II of England, the British throne was inherited by James' Protestant daughter Mary and her Dutch husband, William III and II, of Orange, in the so-called 'Glorious Revolution'. The accession was almost bloodless in England but far from it north of the border, where Jacobite support for the Stuart monarchs was still strong, especially in the Highlands.

A succession of bad harvests from 1695-99, due to exceptionally wet summers and early winters, caused the deaths of thousands from starvation and typhus. To cap it all, the Scottish parliament, the Estates, approved a plan to rival the commercial success of England's East India Company, with a 'Company of Scotland' that would trade with Africa, Asia and America. A far-sighted scheme was hatched by William Paterson (who was also the founder of the Bank of England) to colonise Darien in the isthmus of Panama, as the base for a trading route between the Atlantic and Pacific. Public subscriptions to the shares of *The Scottish Africa and India Company* raised £400,000 from the nobility and the wealthier classes - a sum nearly equal to the entire value of Scottish coinage in circulation. But the Darien venture was a disaster. Two successive waves of colonists were decimated by malaria and attacks by Spanish forces. Interventions by the Royal Navy and the East

India Company were notably absent, stoking Scots' sentiment against King William and the English parliament. The Scottish Company staggered on, but its investors, like many in the general population, were impoverished.

Nevertheless, the 17th century had not been all bad news. As early as 1616 the Scottish Privy Council instructed every Church parish to establish a school, and the Education Acts of 1633 and 1646 imposed a tax on local landowners to provide the necessary funding. The parish schools proliferated, and greatly widened the access to education that was previously the sole preserve of the old grammar or 'High' schools of Glasgow, Edinburgh, Aberdeen, and Dundee, founded many years before. Over the course of the century, rates of literacy in Scotland rose quickly, and were sometimes claimed to be higher than in England and neighbouring countries, though that is debatable. What seems certain is that the Scottish universities were more open and accessible than in England, Germany or France, and the university student body was more representative of ordinary society.

In 1707, the Scottish parliament was dissolved by mutual, and mutually reluctant, agreement with its English counterpart. The union of the parliaments was driven by many complex factors: the anxiousness in England about a renewed drive to full Scottish independence; the desire for a peaceful agreement of a Hanoverian succession to the throne after the death of Queen Anne, who had succeeded the childless William and Mary; and the severely impoverished state of Scotland's nobles and barons. The agreements on finance and Scottish representation at Westminster were seen in Scotland as being much in England's favour. Signing the Act of Union, the Scottish Chancellor, the Earl of Seafield, said wistfully, *"There's ane end of ane auld sang"*. In Edinburgh, the bells of St Giles Cathedral played *'Why am I so sad upon my Wedding Day?'*

Initially, the Act of Union of 1707 proved of considerable detriment to Scotland. Affairs north of the border were at first largely ignored by the new Westminster parliament of 558 MPs, of which the Scottish membership numbered only 45. In the upper house, the ratio of English to Scottish Lords was 196 to 16. Unfavourable English taxes and laws were applied to Scotland, contrary to the spirit of the Act. Popular sentiment in Scotland against the Union, which had been overridden by the decisions of the 'parcel of rogues' in the Estates, was rekindled. The Earl of Mar, one of those who had strongly favoured the Union, informed Queen Anne: *"I think myself obliged in duty to let your Majesty know that... the inclinations and temper of the generality of this country are still as dissatisfied with the Union as ever, and seem mightily soured."* In 1713, the Scottish MPs boycotted parliament after it voted to apply in Scotland the English tax on malted barley. The same year the Earl of Seafield, the ex-Chancellor and previously another strong Unionist, moved in the House of Lords to have the Union dissolved. The motion was just defeated by 71 votes to 67.

In the Highlands, though not only there, Jacobite resentment against the Crown still simmered. When Anne died in 1714, childless despite 17 pregnancies, she was succeeded to the British throne by her second cousin, the Hanoverian, George I. There were disturbances and unrest in Manchester, Somerset, Bath and Oxford. In August 1715, a Westminster Act 'for encouraging Loyalty' required a long list of Jacobite suspects to attend Edinburgh to swear allegiance to the House of Hanover. In Aberdeenshire, Viscount Mar, by now completely disillusioned, organised an armed rebellion together with the Earl of Huntly and George Keith, the 10th Earl Marischal. Their cause was that of the exiled James Edward Stuart, their potential King James VIII and III, nicknamed the 'Old Pretender' by the pro-Hanoverian Whigs. Jacobite forces took Aberdeen, Inverness, Dundee and Perth, and for a while success seemed likely. But Whig-inspired resistance in Glasgow and Edinburgh was strong, and the Jacobite forces were defeated at Sheriffmuir and at Preston. The Jacobite leaders were imprisoned, and two were executed. The Earl of Mar fled from Montrose by sea, and he and Marischal had their estates confiscated by the British government.

Even after the failure of the 'Fifteen' rebellion, the Jacobite claim to the British throne endured and caused more instability and insurrections. The uprising of 1745 in support of Charles Edward Stuart, the 'Young Pretender' known as Bonnie Prince Charlie, came even closer to success. When his forces swept south they advanced as far as Derby before desertions and stretched lines of supply ended their progress. The Battle of Culloden ended in a disastrous, bloody defeat and the 'Forty-Five' marked the final demise of Jacobitism. Charles narrowly escaped with his life, but the sanctions of the British government on the Highland clans were severe and punitive.

For Scottish cities and the Lowlands, the Union was beginning to work. Access to the wider markets of the growing British Empire led to economic growth and a definite rise in the prosperity of the merchant class. The British Navy and Army were major recruiters of Scottish-trained doctors and soldiers. From the schools and universities, a more rational, secular outlook slowly overcame the religious zealotry of the past. Since the Union, Edinburgh was no longer a seat of national government or a real capital city. But she became the nexus of an amazing community of intellectuals, a centre of scientific progress, the leader of a Scottish Enlightenment, and the 'Athens of the North'.

8 The world's youngest professor: Colin Maclaurin (1698-1746)

Illustration 8: Colin Maclaurin by David Steuart Erskine (c. 1795, Scottish National Gallery)

Colin Maclaurin was born in Kilmodan, in Argyllshire, the son of the Reverend John Maclaurin, a Minister in the Church of Scotland, and his wife, Mary Cameron. His father died just a few weeks after his birth, and his mother died before he was nine years old. He was taken into the care of his uncle, the Reverend Daniel Maclaurin, a church minister at the nearby village of Kilfinan.

At the precocious age of 11, Maclaurin was enrolled in the University of Glasgow, to study for the ministry. As well, he soon became interested in geometry, and natural philosophy, no doubt influenced by the professor of mathematics at Glasgow, Robert Simson. Maclaurin graduated with an MA just 3 years later, having written and publicly defended his thesis (in Latin) entitled *'On the power of Gravity and other forces'*. He remained in Glasgow studying for another year, before returning home to Argyllshire where he continued to study both divinity and mathematics.

In 1717, a golden opportunity arose at Marischal College, University of Aberdeen. Mathematics held a specially privileged position at Marischal. Its Liddel Chair of Mathematics had been created in 1613, endowed by Duncan Liddel, an Aberdonian and some-time student of King's College who personified the tendency of Scottish scholars to study and work on the European continent. In the 16th century, Liddel had spent his life teaching mathematics, astronomy and medicine in several German and Polish universities before returning to Scotland in his last years. His bequests to Marischal College were generous. They included property grants to support six students, his personal library of rare books - such as the *De Revolutionibus* and *Commentariolus* of Copernicus - and 6,000 merks (4,200 Scots pounds) for a chair of mathematics. Liddel prescribed in great detail how the bursars and professor should be selected and remunerated, and the subjects for instruction. The chosen professor should demonstrate command of the works of Euclid, Ptolemy, Copernicus, and Archimedes, and were to be paid the handsome sum of 400 merks a year. The Town Council of Aberdeen ordered that the distinguished Liddel scholars should wear black bonnets and black gowns, both in the university and in the town.

The Council struggled for some years to find a suitable candidate for the Liddel Chair, and when they did, the occupants were necessarily embroiled in the political troubles of the 17th century, as occupation of Aberdeen switched between the Covenanting and Royal armies. By 1687 the professorship had become something of a sinecure. Duncan Liddell, namesake and nephew of the original, had acquired the Chair (and an extra 'l' in his name) and proposed to pass it to his son, George, who had proven himself a capable Liddell scholar. George's succession was approved, but two years after his appointment, he was among a group of Aberdonians seized and imprisoned for three weeks by the forces of King William III, as suspected Jacobites. To add insult to injury, in 1706 George Liddell was sacked from the Chair *"frequently being guiltie of scandals and keeping scandalous company together with his present confessione of fornicatione with Jean Bisset"*. He was reinstated after appeal to the King's Advocate, which endeared him not at all to the Town Council.

During the Jacobite rebellion of 1715, widely supported in Aberdeenshire, Marischal College was closed down for two years. The uprising over, George Liddell was sacked again, together with almost all of the professors, the specific charges being that he *"did alwayes frequent the Church during the Rebellion, where the Episcopal Intruders prayed for the Pretender by the name of King James the Eight, did not take the Oaths till after the Rebellion, and has been guilty of such gross immorality as render him of Dangerous and Scandalous Example to the Youth."*

In fact, in 1717, both Aberdeen colleges lost their professors of mathematics. The Regius Professor of Mathematics at King's, Thomas Bower, resigned due to the *'mean and precarious'* nature of his salary. At the time a professor received only a small salary and class fees from individual students, who often obtained extra instruction from private tutors during, and outside, the five months academic session. Many students attended only a few classes. At Marischal, keen to replace the deposed professor Liddell, the Town Council invited candidates to present themselves for a competitive trial at the end of August. Colin Maclaurin and another candidate called Walter Bowman were examined over 10 days by Alexander Burnet, a regent at King's College Aberdeen, and Charles Gregory, Professor of Mathematics at St Andrews (and nephew of James Gregory). Both candidates made a good impression. The examiners reported: *"In the inferior pairts of the Mathematicks ther wes no great odds. Only in 'Euclid' Mr Bowman wes much readier and distincter, and in the last tryall Mr M'Laurine plainly appeared better aquainted with the speculative and higher pairts of the Mathematicks, ... Mr Bowman only hath applyed himselfe to those things that are commonly taught and Mr M'Laurine hath made further advances."*

In awarding the post to the 19-year-old Maclaurin, for an annual salary of 504 Scots pounds, the College had recruited a man who was to be one of the most outstanding mathematicians of his day. Maclaurin's appointment created the world's youngest professor and the record stood until 2008. Maclaurin's ability and enthusiasm for mathematics, both pure and applied, soon showed. He gave a private course to students, in his own chamber, and delivered public lectures. He was responsible for buying apparatus for experimental science, using, with government sanction, surplus money from academic salaries that were withheld during the 'Fifteen' rebellion. In research, he advanced his already considerable findings in geometry, and furthered the ideas of Isaac Newton on curves generated by the intersections of rotating lines. He published early results in two papers in the *Philosophical Transactions*: *'Of the construction and measure of curves'* in 1718 and *'A new method of constructing all kinds of curves'* in 1719.

He expressed delight to have *'an office that allowed me one half of the year to dispose of at my pleasure'*, and took advantage by spending the summers of 1719 and 1721 in London. On his first visit he met Newton, who was immensely impressed by him, and he was elected a Fellow of the Royal Society. For his second visit he travelled, bizarrely, to London in an experimental ship, which was attempting, unsuccessfully, to carry live salmon in its false bottom.

Perhaps stimulated by these contacts and visits, in 1722 he arranged for a substitute to teach his class at Aberdeen, and journeyed to the Continent as the tutor to George Hume, son of Alexander Hume, 2nd Earl of Marchmont, a Scottish judge and politician. During their time in France, Maclaurin wrote an important essay on the mechanics of colliding objects (*'Demonstration des loix du choc des corps'*), which expanded on Newton's laws and won the biennial prize for best paper from the French Academy of Sciences. It was not until the death of his pupil in France, that Maclaurin returned to Aberdeen in December 1724. His reception was frosty. The Aberdeen Town Council was seriously annoyed by his unfulfilled promises of an early return and were not amused by his absence of nearly 3 years *"whereby the students had suffered considerable loss by their not being taught mathematicks as formerly"*. The Council appointed Daniel Gordon, a regent with experience in teaching the subject at St Andrews, to give two hours of instruction per day.

Maclaurin was suspended, *"Untill he should give some reasonable satisfaction and acknowledgement. First, For his going away without Liberty from the Counsell. Second, For his being so long absent from his Charge and not attending the same"*. His defenders pointed to the honour he had brought to the town and college at home and abroad. In April 1725, Maclaurin *"compeared in Counsell and acknowledged that he was sorrie the Magistrats and Counsell had taken offence and promised to be carefull of his Charge hereafter, wherewith the Magistrats and Counsell were satisfied and appointed such Sallarys as were due to the said Professor to be payed to him."*

The young Professor Maclaurin remained at Marischal just long enough to get into an argument with the Principal over his right to vote in Rectorial elections. By November 1725 he was at the University of Edinburgh, appointed Conjunct Professor of Mathematics alongside the aging James Gregory (another nephew of James Gregory of St Andrews). The post at Edinburgh was facilitated for him by Sir Isaac Newton, who offered to pay 20 pounds a year towards his salary. Maclaurin did not bother to inform Aberdeen of his new job. The Council learned of it only *'by the Publict News Prints'*, and indignantly declared his Marischal post vacant in January 1726.

When James Gregory died, Maclaurin solely held the chair in Edinburgh for the rest of his own life. In 1733 he married Anne Stewart, daughter of the Solicitor-General for Scotland, and they had seven children. The city's vibrant intellectual climate fostered many clubs and societies, and Maclaurin led in forming the Edinburgh Philosophical Society, an extension of an existing medical society which developed into the Royal Society of Edinburgh. His numerous published works include a *'Treatise on Fluxions'* (the instantaneous rate of change of a time-varying quantity), which originated as a reply to Bishop Berkeley's criticisms of Newton's calculus. Maclaurin also published a *'Treatise on Algebra'* and *'An Account of Sir Isaac Newton's Philosophical Discoveries'*. Maclaurin is perhaps best known to today's maths students through Maclaurin's Theorem, which shows how a differentiable mathematical function can be expanded as the sum of a series of elements, as a special case of Taylor's Theorem.

MacLaurin's last years coincided with the final Jacobite Rebellion of 1745. He took a leading role in organising the defences of Edinburgh against the Highland army, and when the city fell to the rebels, he fled to York. On the way, suffering from cold and fatigue, he fell from his horse and was injured. His health impaired, he returned home to Edinburgh when the Jacobites had vacated the city, but died shortly afterwards, in June of 1746. His grave lies in the city's Greyfriars Kirkyard.

9 Hume's Fork and Humanism: David Hume (1711 – 1776)

David Hume was not a scientist, doctor or mathematician. But more has been said about Hume and his works than on any other philosopher of the Enlightenment period, and for good reason. His empirical, enquiring, evidence-based approach to philosophical thought underpinned and reinforced much of the scientific endeavour of the Scottish Enlightenment, and he is regarded as perhaps the most important philosopher to write in English. He became renowned also as a historian, economist, and essayist. He was a master stylist in any genre, and his major philosophical works: *'A Treatise of Human Nature'* (1739–1740), *'An Enquiry concerning Human Understanding'* (1748) and *'Concerning the Principles of Morals'* (1751), as well as *'Dialogues concerning Natural Religion'* (published posthumously in 1779), have been widely and deeply influential.

Illustration 9: David Hume by Allan Ramsay (Scottish National Gallery)

Hume's philosophical approach was based on 'empiricism' - that is, the view that knowledge is derived primarily from real-world experience, observation, experiment and our sensory perceptions, rather than purely through thought and pure 'rationality'. He combined this viewpoint with a strong naturalistic rejection of the supernatural and a scathingly sceptical approach to religious belief. Consequently (and appropriately enough!) Hume is often regarded as the first modern Humanist. His conservative contemporaries denounced his writings as works of scepticism and atheism, but his influence is evident in the writings of his close friend Adam Smith, and others. Immanuel Kant reported that Hume's work woke him from his "dogmatic slumbers" and Jeremy Bentham said that reading Hume "caused the scales to fall" from his eyes. Charles Darwin regarded Hume's work as a central influence on his theory of evolution.

Hume was born the second of two sons in Edinburgh in 1711. His family was moderately wealthy, but not rich - his parents were Joseph Home, an advocate, and the Hon. Catherine Falconer, daughter of Sir David Falconer, who was a judge and politician, and sometime Lord President of the Scottish Court of Session. Hume's father died around young David's second birthday, and he was raised by his mother, who never remarried. The family was politically Whiggish and religiously Calvinistic. In childhood David was required to attend the local Church of Scotland, pastored by his uncle.

Educated in boyhood by his widowed mother, he was enrolled in the University of Edinburgh at the age of 11, with the expectation that he would later study Law. His preference soon turned to philosophy and he said that he developed *"an insurmountable aversion to everything but the pursuits of Philosophy and general Learning."* His independence of thought was already apparent, and he did not graduate, writing that *"there is nothing to be learnt from a Professor, which is not to be met with in Books."* Hume left the University at around the age of 15, to study intensively and privately a wide range of subjects. These included the work of that father of the Scottish Enlightenment, Francis Hutcheson, who was soon to be Professor of Moral Philosophy at the University of Glasgow (and the first there to lecture in English rather than Latin). Hutcheson's *'A System of Moral Philosophy'* was influential on both Hume and Adam Smith, though they were to disagree with his conclusions.

Sadly, Hume's intensive studies made him unwell. He came to the verge of a nervous breakdown around 1729, with his doctor diagnosing 'a Disease of the Learned' and prescribing 'a Course of Bitters and Anti-Hysteric Pills' together with a pint of claret every day. Unsurprisingly Hume's mood improved, and he went from being thin and bony to 'sturdy, robust [and] healthful-like', with a developing fondness for cheese and good port.

Hume said he then had *"a very feeble trial"* in the business world, as a clerk for a Bristol sugar importer. His mental crisis had passed, and he remained intent on developing *"a new Scene of Thought"* prompted by his studies. Financially stretched, he moved to France, where he could live cheaply, and settled in the village of La Flèche on the Loire, best known for the Jesuit college where Descartes had studied a century earlier. Hume read French and other continental authors, frequently annoying the Jesuits with his atheistic arguments. In France, between 1734 and 1737, he drafted *'A Treatise of Human Nature'*, subtitled *'An attempt to introduce the experimental method of reasoning'*, his first major work.

The *Treatise* explored several topics such as space, time, causality, external objects, the passions, free will, and morality, offering original and often sceptical assessments of these notions. Hume had been greatly impressed by Isaac Newton's achievements in physical science, and sought to introduce the same methods of reasoning, based on evidence and experimentation, into the 'Study of Man', and to discover the *"extent and force of human understanding"*. Arguing against philosophical 'rationalism', he introduced the famous 'problem of induction', arguing that inductive reasoning and our beliefs regarding cause and effect cannot be justified by intuitive reasoning alone. Instead, our faith in induction and causation is the result of observation and customary, practical experience: *"Custom is the great guide of human life"*. Hume also introduced an important division between 'relations of ideas' and 'matters of fact and real existence'. Another way to express this is his division between: (a) ideas which are necessarily true from the rules of analytical logic or language - e.g. all bachelors are unmarried men, and

(b) statements about the world which are contingent and dependent on our perceptions and experience - e.g. all bachelors are extremely rich - which may be probable or possible, and may or may not be true. This dichotomy has become a fundamental idea in Western philosophy and became known as Hume's Fork.

Hume moved to England in 1737, and set about publishing the Treatise. To make it less contentious and to avoid the hostility of such as the religious philosopher Joseph Butler, Bishop of Durham, he *"castrated"* his manuscript, deleting his controversial discussion of miracles, along with other *"nobler parts"*. Books I and II were published anonymously in two volumes in 1739 and Book III appeared the next year. Although now regarded as a philosophical masterwork, the *Treatise* met a poor reception. Hume famously said, in his later autobiography, that *"it fell dead-born from the press, without reaching such distinction, as even to excite a murmur among the zealots. But being naturally of a cheerful and sanguine temper, I soon recovered from the blow and prosecuted with great ardour my studies in the country."* In 1741 and 1742 Hume was more successful in publishing his two-volume *'Essays, Moral and Political'*, which was written in a more popular style.

Emboldened, he applied in 1744 for the Chair of Ethics and Pneumatical (Mental) Philosophy at the University of Edinburgh. Opponents condemned Hume by citing his critiques of religion, and unfortunately, chief among them was the Principal of the University, the clergyman William Wishart. Lists of allegedly dangerous propositions from Hume's *Treatise* circulated, some rumoured to have been penned by Wishart himself. In fact Hume's self-censored critique of religion in the *Treatise* was fairly mild. He had written: *"In matters of religion, men find pleasure in being terrified and...there are no more popular preachers than those who excite the greatest sadness and the most gloomy passions"* and *"Generally speaking, the errors in religion are dangerous; those in philosophy only ridiculous."* In the face of Wishart's opposition, the Edinburgh Town Council consulted the Edinburgh ministers, of whom 12 of 15 voted against Hume, who quickly withdrew his candidacy.

In 1745 the second Jacobite rebellion resulted in the temporary occupation of Edinburgh by the Highland clansmen supporting the claim of 'Bonnie Prince' Charles Stuart to the British throne. Hume was away - during the debacle of his Edinburgh professorship application, Hume had taken a position in England as a tutor to a young nobleman, the Marquis of Annandale, only to discover that his pupil was insane. He swiftly accepted an invitation from his cousin, Lt-General James St. Clair, soldier and Member of the British Parliament, to join him as his Secretary. The result was several years of foreign adventures, accompanying St. Clair on military expeditions in Canada and France, and to embassy posts in the courts of Vienna and Turin.

Hume continued to write and *'An Enquiry concerning Human Understanding'* appeared in 1748. It restated the central ideas of Book I of the *Treatise* (Hume's Fork) and the discussion of liberty and necessity from Book II. He also included controversial material he had excised from the Treatise, and hardened his stance on scepticism and empiricism. In the final line of the *Enquiry* he asks us to enquire of any book, *"Does it contain any abstract reasoning concerning quantity of number? No. Does it contain any experimental reasoning concerning matter of fact and existence? No. Commit it then to the flames: For it can contain nothing but sophistry and illusion."*

In Section X, *'Of Miracles'*, Hume is bold in his challenge to Christian orthodoxy. It begins: *"...the authority, either of the scripture or of tradition, is founded merely in the testimony of the apostles, who were eye-witnesses to those miracles of our Saviour, by which he proved his divine mission. Our evidence, then, for the truth of the Christian religion is less than the evidence for the truth of our senses; because, even in the first authors of our religion, it was no greater; and it is evident it must diminish in passing from them to their disciples; nor can any one rest such confidence in their testimony, as in the immediate object of his senses.... A wise man, therefore, proportions his belief to the evidence'.*

And later, even more explicitly: *"A miracle is a violation of the laws of nature; and as a firm and unalterable experience has established these laws, the proof against a miracle, from the very nature of the fact, is as entire as any argument from experience can possibly be imagined. Why is it more than probable, that all men must die; that lead cannot, of itself, remain suspended in the air; that fire consumes wood, and is extinguished by water; unless it be, that these events are found agreeable to the laws of nature, and there is required a violation of these laws, or in other words, a miracle to prevent them? Nothing is esteemed a miracle, if it ever happen in the common course of nature. It is no miracle that a man, seemingly in good health, should die on a sudden: because such a kind of death, though more unusual than any other, has yet been frequently observed to happen. But it is a miracle, that a dead man should come to life; because that has never been observed, in any age or country."*

One outcome of the publication of the *Enquiry* was a charge of heresy, against which Hume's successful defence was to restate his atheism. Arguably another consequence was the failure of Hume's application for the Chair of Logic at the University of Glasgow. Again his application was rejected. Even his friend Adam Smith, who had previously held the same professorship, had advised him against an application. Hume never did hold an academic post, but took a job as Librarian of the Advocate's Library in Edinburgh, which at least gave him the chance to pursue his interests in history. There, he wrote most of his six-volume '*History of England (and Great Britain)*' which was published over the period 1754 to 1762. The first volume was initially badly received, partly for its defence of Charles I, and partly for its attacks on Christianity, but the *History* became a commercial success - a standard and much-cited work which gave Hume the financial independence he sought. His publisher pleaded for another volume and Hume's typically whimsical response was, *"I have four reasons for not writing: I am too old, too fat, too lazy and too rich."*

Back in 1751, the first *Enquiry* had been joined by a second, '*An Enquiry concerning the Principles of Morals*'. It was a substantially rewritten version of Book III of the *Treatise*, and Hume regarded it as *"incomparably the best"* of all his works. Another publication, the '*Political Discourses*', appeared in 1752, and soon Hume was drafting his '*Natural History of Religion*' and '*Dialogues concerning Natural Religion*', which would eventually be published posthumously, since Hume had been persuaded, by friends and publishers, to suppress any more controversial writings on religion during his lifetime.

Between 1763 and 1765 Hume was the Private Secretary to Lord Hertford, the British Ambassador to France, where he met Voltaire and Rousseau. In 1767 he held a government post in London, as Under Secretary of State for the Northern Department. In 1768 he returned to a merry and sociable life in a house he had built in the Georgian New Town of Edinburgh. It was here that he recounted a tale. Going home by crossing the bog (left by the draining of the Nor'Loch to make way for the New Town), he slipped and fell from the path into the bog. Stuck, he called for help as twilight fell. An Edinburgh fishwife approached, but looking down, recognised him as 'Hume the Atheist' and refused to help. Hume pleaded, and asked if her religion did not require her to do good, even to enemies. She replied, *"That may well be - but ye shall na' get oot o' that till ye become a Christian yersel' and repeat the Lord's Prayer and the Belief"* (meaning the Apostolic Creed). To her astonishment Hume did exactly that, and the old woman reached down and pulled him out.

In 1776, at age 65, Hume died from an intestinal cancer after being ill for many months. He was unmarried, though he had formed a close relationship with an attractive, vivacious, and intelligent woman in her twenties, called Nancy Orde, who was the daughter of Baron Orde of the Scottish Exchequer. She was described as "one of the most agreeable and accomplished women" and renowned for her cheeky sense of humour. She chalked 'St. David's Street' on the side of Hume's house and the street still bears that name today. The couple were close enough that it was rumoured that they were engaged. Hume added a codicil to his will, which included the gift to Nancy of *"Ten Guineas to buy a Ring, as a Memorial of my Friendship and Attachment to so amiable and accomplished a Person".*

Hume's influence on the thinking of philosophers and scientists has been vast. Kant, Bentham and Darwin have already been quoted. Arthur Schopenhauer said "there is more to be learned from each page of

David Hume than from the collected philosophical works of Hegel, Herbart and Schleiermacher taken together". Albert Einstein, in 1915, wrote that he was inspired by Hume's 'logical positivism' when formulating his own theory of special relativity. Karl Popper wrote: *"Knowledge ... is objective; and it is [either] hypothetical or conjectural. This way of looking at the problem made it possible for me to reformulate Hume's problem of induction"* and *"I approached the problem of induction through Hume. Hume, I felt, was perfectly right in pointing out that induction cannot be logically justified."*

Hume died with dignity and humour. At a last dinner, Adam Smith complained of the cruelty of the world for taking Hume from them. Hume said: *"No, no. Here am I, who have written on all sorts of subjects calculated to excite hostility, moral, political, and religious, and yet I have no enemies; except, indeed, all the Whigs, all the Tories, and all the Christians."* On Hume's death, Smith wrote that Hume approached *"as nearly to the idea of a perfectly wise and virtuous man, as perhaps the nature of human frailty will permit."*

10 Founding the science of economics: Adam Smith (1723-1790)

Illustration 10: Adam Smith, after James Tassie (c.1787)

Is economics truly a science? The Scottish polymath Thomas Carlyle called it 'the dismal science', and old jokes about economists persist and abound. For example: *"The First Law of Economics: For every economist, there exists an equal and opposite economist. The Second Law: They're both wrong."* And *"There are three types of economist: those who can count and those who can't."* And so on. But whatever the scientific status of economics, there is no doubt that the central principles of modern economic thought stem from the work of Adam Smith, a giant figure of the Scottish Enlightenment, widely regarded as the 'Father of Economics'. His key work *'An Inquiry into The Nature and Causes of the Wealth of Nations'* has hugely influenced subsequent economic theory and practice.

Before Smith, a nation's wealth was measured by its stock of gold and silver. Imports from abroad were seen as damaging - since the nation's wealth must be used to pay for them. In contrast, exports were seen as good - because exports produced a return flow of the precious metals. Controls were used to prevent the drain of the nation's metal wealth: import tariffs and taxes; incentives and subsidies to exporters;

and political acts to protect domestic industries. This kind of protectionism was used within nations too. Cities prevented the inward migration of workers and tradespeople from other towns. Governments were petitioned by producers and merchants to provide protective monopolies. Labour-saving devices were opposed as a threat to existing practices.

Smith's work exposed the mistakes of this thinking. He showed that in free exchange of goods and services, both sides benefit, pointing out that nobody trades in order to lose, and both buyer and seller profit from the trade. Imports are as valuable to us as our exports are to our trading partner. And because trade benefits both sides, trade increases prosperity just as well as agriculture or manufacture. A nation's wealth is not measured by the gold and silver in its vaults, but by the sum of its production and trade – in modern terms, the gross national product.

These insights sprang from Smith's work in moral philosophy as well as economics. Influenced by his mentor Francis Hutcheson and close friend David Hume, Smith regarded freedom and people's self-interest as an 'invisible hand' that guides society to co-operation and order, with resources steered automatically to the most highly valued objectives.

Smith was born in Kirkcaldy, Fife, and his mother was Margaret Douglas, daughter of the landowner Robert Douglas. His father, also Adam Smith, was a lawyer and customs controller who died 2 months before his son's birth. The young Adam attended the excellent local Burgh School, studying the usual curriculum of Latin, History and Mathematics. He was enrolled at the University of Glasgow aged 14 and studied moral philosophy under the charismatic professor Francis Hutcheson who instilled a love of reason, moral responsibility, and liberal values. On graduation, Smith was awarded a Snell fellowship to Balliol College, Oxford. He was not impressed, writing: *"In the University of Oxford, the greater part of the public professors have, for these many years, given up altogether even the pretence of teaching"* and complained that on being seen reading a copy of Hume's *'A Treatise on Human Nature'*, he was punished and the copy was confiscated. He left Oxford, stressed and unhappy, in 1746, before the end of his scholarship, and returned to private study.

He began to give public lectures at the University of Edinburgh around 1748, supported by Henry Home, Lord Kames, a judge and a driving-force of the Edinburgh Philosophical Society. Smith was no great lecturer, but his reputation grew among the learned societies of Edinburgh. He met David Hume in 1750 and forged a warm personal and professional friendship, with frequent correspondence, that was to last until Hume's death 26 years later. Smith's growing reputation saw him appointed to the Chair of Logic at the University of Glasgow, moving the following year to the Chair of Moral Philosophy, the professorship previously held by Hutcheson. Thus began the happiest and most productive period of his life.

Smith's reputation was further enhanced by his publication in 1759 of *'The Theory of Moral Sentiments'*, based largely on his lectures at Glasgow, where attendance was swelled by wealthy students from home and abroad. *'Moral Sentiments'* is an attempt to explain the source of our ability to form moral judgements, assuming that we begin life with no moral sense at all - in distinction to Hutcheson's assumption. Smith proposed that the act of observing others, and the judgements they form, creates our own self-awareness and concern on how others judge our behaviour. This feedback from the judgements of others, plus our judgements of them, creates a 'mutual sympathy of sentiments' which is the basis of our conscience and morality. Smith was hugely foresighted and anticipated the globalisation of trade and industry that we have experienced today, with all its attendant benefits and hazards of interdependency. He predicted the rising economic might of Asia, and in *The Theory of Moral Sentiments'* he wrote : *"Let us suppose that the great empire of China, with all its myriads of inhabitants, was suddenly swallowed up by an earthquake, and let us consider how a man of humanity in Europe, who had no sort of connection with that part of the world, would be affected upon receiving intelligence of this dreadful calamity. He would, I imagine, first of all, express very strongly his sorrow for the misfortune of that unhappy people, he would make many melancholy reflections upon the precariousness of human life, and the vanity of all the labours of man, which could thus be annihilated in a moment. He would too, perhaps, if he was a man of speculation, enter into many reasonings concerning the effects which this disaster might produce upon the commerce of Europe, and the trade and business of the world in general."*

Over the next years, Smith intensified his focus on economics and the theory of law, while perhaps losing interest in moral philosophy and his professorship. He resigned his Chair at Glasgow in 1764 to take a private job tutoring Henry Scott, the young Duke of Buccleuch. The attraction no doubt was the remuneration of £300 per year (plus expenses) along with a £300 per year pension, which was roughly twice his income at the University. It was said that he offered to return the fees he had collected from his former students, since he resigned halfway through their term; but the students refused.

The new job involved travelling as a tutor in Europe, mainly in France and Switzerland. They spent 18 months in Toulouse which Smith found to be boring, writing to Hume that he *"had begun to write a book to pass away the time."* The group next visited Geneva, where Smith met Voltaire, then aged 70. They next moved to Paris where Smith met Benjamin Franklin, a Europhile as well as the key figure of the American Enlightenment. In Paris Smith made another contact who shaped his thinking. François Quesnay was the founder of a school of thought named *'Physiocracy'* (from the Greek for 'government of nature') which opposed mercantilism, the then-prevailing economic policy that sought to maximise exports while minimising imports. The Physiocrats' slogan was *'Laissez faire et laissez passer, le monde va de lui même!'* (Let do and let pass, the world goes on by itself!), and their emphasis was on promoting 'productive work' as the main source of national wealth.

Smith, now aged 43, returned to Scotland in 1766 and settled again in Kirkcaldy to think, study and write during the next decade. The result was the publication of his masterwork *'An Inquiry into The Nature and Causes of the Wealth of Nations'* which was published in 1776. Smith had maintained his network of intellectual friends and had been elected a

Fellow of the Royal Society, which no doubt helped *The Wealth of Nations'* to be an instant publishing success. The first edition sold out in 6 months. It used real-world examples to illustrate Smith's themes. In his example of a pin factory, Smith showed how specialisation can boost productivity. By specialising, people can improve their skills, and can employ labour-saving machinery to boost production. Then the exchange of those specialist products spreads the benefits of specialisation across the whole population. And although often cited as promoting unrestricted free enterprise, Smith's writing is more nuanced, influenced by his moral concerns to limit inequalities and to balance the power of 'Masters' with the rights of workers. *"We rarely hear, it has been said, of the combinations of masters, though frequently of those of workmen. But whoever imagines, upon this account, that masters rarely combine, is as ignorant of the world as of the subject. Masters are always and everywhere in a sort of tacit, but constant and uniform, combination, not to raise the wages of labour above their actual rate...Masters, too, sometimes enter into particular combinations to sink the wages of labour even below this rate. These are always conducted with the utmost silence and secrecy till the moment of execution; and when the workmen yield, as they sometimes do without resistance, though severely felt by them, they are never heard of by other people".*

In contrast, Smith wrote, when workers unite, *"the masters...never cease to call aloud for the assistance of the civil magistrate, and the rigorous execution of those laws which have been enacted with so much severity against the combination of servants, labourers, and journeymen."* Smith's liberal inheritance from Hutcheson, and the influence of Hume, meant that he also supported the imposition of progressive 'fair' taxes which took proportionately more from the rich than poor, and he was an opponent of imperialism, slavery and extreme inequality of wealth.

Smith moved to Edinburgh in 1778, never married, and was chronically untidy and absent-minded. He had a funny walk and a speech impediment. He often talked to himself, and poked fun at his own appearance *"I am a Beau in nothing but my books".* He was reluctant to sit for portraits. According to one story, on a tour of a tanning factory, and while discussing free trade, Smith walked into a huge tanning pit and needed help to get out. On another occasion he was said to have walked, in deep reverie, for 15 miles in his nightgown before realising his whereabouts.

Adam Smith died, aged 67, at Panmure House in Edinburgh, six years after the death of his mother. Shortly before his death he had nearly all his manuscripts destroyed, but the legacy influence of his published work is wide and profound. *'The Wealth of Nations'* has become hugely influential and created the subject of political economics as a systematic discipline. For the western world, it was the most significant book on the subject ever published.

11 Anatomy, obstetrics and scientific surgery: William and John Hunter (1718-1793)

The Hunter brothers were born, 10 years apart, at Long Calderwood, today part of the new town of East Kilbride in Lanarkshire. Their pioneering exploits in anatomy and surgery were to be admired and revered, but were also the subject of lurid stories and scurrilous rumours.

The brothers' family background was fairly ordinary. Their father, John, was a retired grain merchant and their mother, Agnes Paul, was the daughter of a Glasgow magistrate. The couple had ten children - of whom only William, John, James and Dorothea survived to adulthood. Family income was meagre, but Agnes was well-educated and schooled her children in the theatre, literature and the arts. William, the 7th child, enrolled at age 13 at the University of Glasgow, where Francis Hutcheson was influential on him, and he studied a classical curriculum for five years.

Illustration 11: William Hunter by Allan Ramsey (Hunterian Museum and Art Gallery)

William turned to a medical career at the age of 19 in 1737, and his subsequent career was heavily influenced by several of the many eminent Scottish physicians of the Enlightenment. He was first apprenticed to William Cullen, their well-regarded family doctor in neighbouring Hamilton. William was Cullen's resident pupil for the next three years and they proposed to enter into partnership. Cullen was a life-long influence on William, and on the Scottish Enlightenment generally. Cullen became personal physician to the Duke of Hamilton, and in 1751 he was appointed Professor of the Practice of Medicine at the University of Glasgow. Later he was recruited by Lord Kames to become Professor of Chemistry and Medicine at the Edinburgh Medical School, the oldest school of medicine in Britain and one of the oldest anywhere, and at that time the leading centre of medical education in the English-speaking world. As well as William Hunter, Cullen tutored many students who would have high-achieving careers including Gilbert Blane and the chemist Joseph Black, and he became personal physician to David Hume and indeed the King (in Scotland). Cullen combined superb patient-facing skills with a flair for teaching and a keen intelligence, which he applied to seminal work on the classification of diseases, and a systematic approach to organising medical knowledge. William Hunter referred to him warmly: *"to whom I owe more and more love of all the men of the world"*. Cullen would describe William more reservedly as *"a young man whose conversation was remarkably vivacious and pleasant, and all his behavior was more strict and constantly correct than that of any other"*.

William Hunter had other important influences. As a student at Edinburgh, he studied anatomy under Professor Alexander Monro *primus,* one of the founders of the Edinburgh Medical School. In 1740 Hunter moved to London, to begin studies in midwifery as the pupil of the highly innovative William Smellie, another close friend of Cullen. Smellie, practicing at the recently-founded St. George's Hospital, was the most eminent 'man-midwife' or 'accoucheur' in London, and was a forward thinker in what we would call today 'obstetrics' (a term not coined until much later).

At this time, Hunter was also introduced to the eminent physician and anatomist James Douglas, another Edinburgh man and a Fellow of the Royal Society for the past 35 years. William became anatomy assistant to Douglas and tutor to his son, also James. Hunter was also generously given accommodation in the Douglas household. When Douglas died in 1742, aged 67, he insisted that his family promise to send his son and William to Paris, to make further studies in anatomy and surgery. Hunter, keen to take advantage, first studied in Paris in 1743, where he absorbed the best available courses on anatomy and surgery, and was able to dissect a whole body for the first time. This convinced him that the practical experience of dissection was essential in the accurate teaching of anatomy and surgery. William returned to London in 1744, at the age of 26, having completed a first-class, wide-ranging medical education. He quickly became an eminent teacher and doctor at London's Middlesex Hospital, The British Lying-in Hospital for Married Women, and in private practice at Covent Garden. He was admitted to the Surgeons' Corporation, and presented his first paper to the Royal Society, *'On the structure and diseases of articulating cartilages'*.

Illustration 12: John Hunter by Sir Joshua Reynolds (Wellcome Collection)

In 1748 William's youngest sibling, John, 21 years old, joined him in London to learn dissection. John had been no scholar, and had no academic training, but was a young man of impatient curiosity, practical intelligence and manual dexterity. By the next year John was supervising the dissection class in William's anatomy school, and had another responsibility: the procurement of corpses. The lack of available corpses was the main limitation for the teaching of anatomy in Britain, and it was a difficult business. Nevertheless, soon the anatomy school could provide a corpse for each of its growing population of students. This raised rumours and controversy at the time and since. What was the source of the bodies? Grave robbing was a criminal offence, but that failed to deter those operating what was a lucrative trade. The Hunters were discreet but were unable to avoid attracting attention, and rumours swirled around William for many years. A newspaper commented on the high number of bodies that were delivered to his residence (and anatomy school) at Great Windmill Street and led to angry mobs demonstrating outside his house. A cartoon engraving by William Austin in 1773 depicts a night-watchman disturbing a body-snatcher who has dropped the stolen corpse he had been carrying, while William Hunter runs away.

It has even been suggested that the Hunter brothers were responsible in some way for the deaths of many women whose corpses were used for their studies on pregnancy. In fact, there is no evidence for this. The rate of death of pregnant women in London during these years was high, given the prevalence of conditions like pre-eclampsia, for which no treatment was known at the time. The act of childbirth itself was hazardous due to the unhygienic - and sometimes filthy - conditions in hospital wards, which led to fatal infections, as proved decades later by the work of Ignaz Semmelweis in Vienna. William Hunter's approach to 'man-midwifery' anticipated this. For example, he disliked and minimised the use of forceps in delivery.

William shared the contemporary enthusiasm for meeting clubs, coffee houses and taverns. There he cultivated a range of eminent friends and contacts, including the artist Allan Ramsay, the writer Tobias Smollett, and the engraver William Strange, all of whom he had known in Scotland, as well as Samuel Johnson, and Royal Society contacts Sir Joshua Reynolds and Benjamin Franklin. He would arrive at his clubs after his lectures had finished, at around 9 pm, and relax with fellow countrymen and friends over a glass of claret. His usual toast was *'May no English Nobleman venture out of the World without a Scottish Physician, as I am sure there are none who venture in'*. In 1762, he was asked to attend the birth of the future King George IV, though it was a midwife who actually delivered Prince George. William was appointed Physician Extraordinary to Queen Charlotte in 1764 and until 1783 supervised the delivery of her 14 other children.

Aided by careful dissections, the Hunter brothers collaborated in studies of the pregnant womb and the placental circulation, and their findings were often described in William's lectures. They had shared the house at Covent Garden until 1755 when William moved to a larger house and left John in charge of the anatomy school. Around 1760 they had a major falling out, fuelled by John's perceived lack of accreditation on the placental work in particular. Some of this work featured in William's most famous book *The Anatomy of the Human Gravid Uterus'*, published in 1774 and containing beautifully detailed illustrations by the medical artist Jan van Rymsdyk.

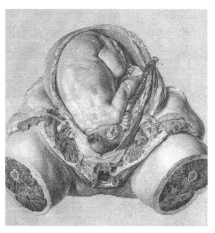

After the disagreement with William, John went his separate way, although his first published paper, *'The State of the Testis in the Foetus and on the Hernia Congenita'*, appeared in 1762 within William's *'Medical Commentaries'*, with more illustrations by van Rymsdyk. John took a

Illustration 13: Anatomy of late pregnancy (Jan van Rymsdyk, from 'The Anatomy of the Gravid Uterus' by William Hunter)

commission as an army surgeon and served for three years during which he objected to prevailing surgical practices, such as the dilation of gunshot wounds. He took the opportunity of campaigns in the Seven Years War in France and Portugal to collect natural history specimens including lizards, eels and fossils. On leaving the army, after the Peace of Paris in 1763, he practiced privately, including a collaboration with James Spence, a London dentist, to conduct some of the first viable tooth transplants between living people.

John became a Fellow of the Royal Society in 1767, and his fame as a teacher and as a methodical, talented surgeon was growing. There were mis-steps too. He became interested in venereal disease, but wrongly hypothesised that two diseases could not exist simultaneously in the same organ. Therefore, he believed, syphilis and gonorrhoea were different symptoms of the same sexual illness. In keeping with his empiricist mantra *'Don't think, try; be patient, be accurate...'* he devised an experiment to inoculate a patient with pus from a gonorrhoeal lesion - the patient being reputed to be himself. Unfortunately, the sample was contaminated with syphilis, and both diseases developed. Hunter reached the wrong conclusion and took this as proof of the hypothesis that the two diseases were the same - a rare example of him clinging to prejudice in the face of contrary evidence, and a mistake which set back understanding of venereal diseases for many years. The outcome of the experiment delayed his marriage by three years while the diseases were cured. Nevertheless in 1771 John, then 43, married the 29-year-old Anne Home, daughter of a Scottish military surgeon, Robert Boyne Home, and sister of Everard Home, who declined a scholarship at Trinity College Cambridge to become John Hunter's pupil. Everard became Surgeon General to King George III and was knighted in 1813, but he is now suspected of plagiarising much of Hunter's work, and of destroying legacy papers to conceal the deception. In any event, the marriage of John and Anne was a surprise to many, for their interests and tastes were greatly different. Anne was a charming and accomplished socialite who, in due course, gave birth to four children, though only two survived infancy. She was tall, blonde, skilled in the harpsichord, painting and poetry, and a noted hostess of cultured 'bluestocking' women's meetings. John was dishevelled, poorly read and absorbed in his work. During their years of marriage Anne uncomplainingly endured a chaotic house and workplace stuffed with mummified exotica, skeletons, fossils and cadavers. Somehow she managed to host one of the liveliest salons in London, with regular, sophisticated soirées.

The stability of John Hunter's marriage helped his career too. His fame grew with his surgical practice and his private practice and hospital duties occupied much of his day. His evenings were often spent at meetings of learned societies, or in writing notes upon his cases or subjects of research. He performed many surgical operations of great difficulty with unprecedented success and meticulous care, and has become known as the founder of 'scientific' surgery. In 1783 he and Anne moved to a large house, bringing with them his collection of 14,000 specimens of plants, fossils and live animals. Soon thereafter in a controversial competition with other collectors, he paid for the skeleton of the Irish 'giant' Charles Byrne, who was 7'7" (2.31m) tall. This was clearly against Byrne's deathbed wishes since he had asked to be buried at sea. Hunter published a detailed study of the skeleton and anatomy, but the ethics of the acquisition were dubious at best.

In 1786, John was appointed Deputy Surgeon to the British Army and in March 1790, he was made Surgeon General by the Prime Minister, William Pitt (the Younger). In this post, he reformed the system for the selection of army surgeons to be based on experience and merit, rather than on the patronage and nepotism that had previously held sway.

John had been on bad terms with his William since they argued in 1760, and there was little dialogue in later years. In 1780 their disputes about discoveries connected with the placenta and uterus led to a final schism. William continued to practise, and received many more honours at home and abroad. In March 1783, suffering severe illness for several days he gave his introductory lecture on the operations of surgery, but fainted near the end, and was carried to bed. He died that month, aged 64. He had never married.

John could be hot-tempered and readily provoked, and his former student, Edward Jenner (later developer of the smallpox vaccination) had diagnosed John with angina. Hunter's attacks were frequently precipitated by emotional upsets and he predicted correctly that one of these would cause his death, saying *"My life is in the hands of any rascal who chooses to annoy or tease me."* His death in 1793 was indeed due to a heart attack, brought on by arguments at the board of governors of St George's Hospital, concerning the appointment of his successor and the admission of students. He is buried in Westminster Abbey. His widow Anne was left poorly provided for, until in 1799 she was given £15,000 by Parliament for John's collection of books, paintings, instruments and anatomical specimens.

Both the Hunter brothers left legacies of great medical advancements as well as considerable collections of scientific exhibits and specimens. The Hunterian Museum and Art Gallery in the University of Glasgow houses William's huge collection, which goes beyond medicine and anatomy and includes books, manuscripts, prints, coins, shells, zoological specimens, and minerals. John Hunter's collection was given to the Royal College of Surgeons of England, and several thousand items from his collection can be seen at their Hunterian Museum in Lincoln Fields, London. In East Kilbride, the achievements of the brothers are remembered in the names of Hunter Primary School, Hunter High School, and in their birthplace, Hunter House.

12 Lemons, Limeys, and Scots against scurvy: James Lind, Gilbert Blane and Lauchlan Rose (1716-1885)

Commander George Anson, a hero of the British Royal Navy, returned home in 1744 having sailed around the world on a four-year voyage of discovery and conquest. His fleet had captured a Spanish galleon carrying nearly 36,000 ounces of silver and more than a million 'pieces of eight'. He and his men paraded triumphantly in London displaying thirty-two wagons filled with the bullion. But his crew had paid a very high price. Two-thirds of his sailors - more than 1000 men - had been killed by typhus and by the disease known as 'scurvy'.

Illustration 14: James Lind by Sir George Chalmers (James Lind Library)

Scurvy was an ever-present hazard which had plagued seafarers for centuries, in voyages of any length beyond a few weeks. First recorded by Vasco de Gama, its prevalence and incidence grew with increased seaborne trade and lengthening voyages of exploration. It would eventually kill an estimated two million sailors. Victims developed swollen, spongy and bleeding gums, loose teeth, and bulging eyes.

Severe and easy bruising with bleeding into the skin was combined with scaly, dry, brownish skin, and very dry hair that curled and broke off close to the skin. The symptoms were accompanied by severe pains in joints and muscles, causing the sufferers to adopt frog-leg postures to ease the agony. Scurvy patients disintegrate slowly and die painfully. And the disease was routinely killing more sailors than any other cause.

We know now that scurvy is caused by vitamin deficiency, specifically lack of vitamin C. In the 18th and 19th centuries the concept of vitamins was unknown, and various causes were suspected. Supposed remedies included bloodletting, a dose of sulphuric or hydrochloric acid, vinegar, or mercury paste. It was also suspected that citrus fruit provided protection, and a British East India Company surgeon called John Woodall had recommended this as a solution, but his views were ignored. The idea that eating fruit could be a cure for anything was regarded as faintly ridiculous.

Two years after Anson's return, a 30-year-old Scottish naval surgeon named James Lind joined *HMS Salisbury* on a voyage to patrol the English Channel and the Bay of Biscay. Within weeks, one in ten of the sailors were showing signs of scurvy. Lind decided to test alternative remedies to determine the most effective treatment, and thus designed the world's first controlled clinical trial.

Lind was born to his merchant father James and mother Margaret in Edinburgh in 1716. He was educated at the Edinburgh High School and the Edinburgh University Medical School, where he studied anatomy under Alexander Monro *primus*. He was awarded his MD degree in 1748, aged 32, having written his thesis on venereal disease. By then he had served, without formal qualifications, for 10 years as a naval surgeon, seeing service in the Mediterranean, Guinea and the West Indies.

In the Spring of 1747, in the Bay of Biscay aboard the *Salisbury*, Lind designed his clinical trial by first identifying 12 sailors with similar symptoms of scurvy: *"the putrid gums, the spots and lassitude, with weakness of their knees"*. He positioned their hammocks in the same part of the ship and ensured that they received the same food, to create a fair basis for the trial. *"They lay together in one place, being a proper apartment for the sick in the fore-hold; and had one diet common to all, viz. water gruel sweetened with sugar in the morning; fresh mutton-broth often times for dinner; at other times light puddings, boiled biscuit with sugar, etc., and for supper, barley and raisins, rice and currants, sago and wine or the like."* He divided the patients into 6 pairs, and treated each pair with a different daily regimen, either: a quart of cider; 25 drops of elixir of vitriol (sulphuric acid) three times; 2 spoonfuls of vinegar, three times; a half-pint of seawater; a paste of garlic, mustard, radish and myrrh; or 2 oranges and a lemon. Cleverly, another group with the symptoms were also observed, and kept on normal naval rations. They provided, in today's terms, the control group.

The supply of the fruit ran out after only 6 days, so the planned trial length of 14 days was not completed, but even so the results were definitive. The two sailors consuming the oranges and lemons had almost completely recovered, and one of them was fit for duty. Of the other groups, all were still sick, apart from the cider drinkers who showed slight signs of improvement.

Lind was convinced, while of course having no idea that the key ingredient of the cure was vitamin C. But he was an unassuming and diffident man, and failed to publicise the results of his research until 6 years later when he published *'A Treatise of the Scurvy'* in 1753. In it he was conscious of challenging accepted 'wisdom' and wrote *"... it is no easy matter to root out old prejudices, or to overturn opinions established by time, custom and great authorities... Indeed, before this subject could be set in a clear and proper light, it was necessary to remove a great deal of rubbish... Where I have been necessarily led, in this disagreeable part of the work, to criticise the sentiments of eminent and learned authors, I have not done it with a malignant view of depreciating their labours, or their names; but from a regard to truth, and to the good of mankind."* The book was dedicated to Commander Anson, but it was 400 detailed pages long and attracted little attention or endorsement. Lind's actions after his seaborne tests had not helped. He developed a concentrated lemon juice syrup (called 'rob') designed to be easy to transport, store and administer. The syrup was produced by heating and evaporating lemon juice, which, unknown to Lind, destroyed the vitamin C. This rendered the concoction completely ineffective, as confirmed on Captain James Cook's first Pacific voyage, who used mainly sauerkraut as the dietary antiscorbutic for his men. The simple solution of citrus fruit was not adopted and Lind himself went on to recommend other remedies including a dietary supplement of watercress grown on board ships, which was actually practiced for a while.

James Lind retired from the Navy just a year after his historic and ingenious clinical trial, and spent the next 25 years travelling, publishing and working as chief physician the Royal Haslar Hospital in Gosport, where he continued to research and propose methods to improve the health and hygiene of seafarers. He discovered that the steam of heated salt water was fresh, and suggested the use of solar energy to distil seawater, though seawater distillation on a useful scale became practicable only in 1810 when a new type of naval cooking stove was introduced. Lind's inventiveness ranged also to the design of ingenious electrical machines and heating devices, which he wrote about to his friend and frequent correspondent James Watt, who he had known from Scotland. In one example, written from Edinburgh to Watt in October 1768 " ....*my last works are a model of a cylindric Chinese bellows with a forge, and a pocket Kittchen containing a pan, brandy flask & cup, butter dish, salt, pepper, soy, mustard, and a furnace, it is capable in a few minutes to shrew a dish of stakes, make a hash or dress a fowl and the whole is only 7 inches long, 5 broad and 2½ thick. It carries above an Eng. Pint of spirits, ¼lb of butter the ....near the size of....which is the furnace...*"

However, beyond any doubt Lind is best remembered for his world first: the first controlled clinical trial, for which he is regarded as the first truly modern clinical investigator. To celebrate the day that he started his trial to find an effective treatment for scurvy, 'International Clinical Trials Day' is celebrated around the world on or near the 20th of May each year. The James Lind Institute, based in India, continues his work in clinical education and research.

Lind was elected a Fellow of the Royal Society of Edinburgh in 1783, and made a Knight Commander of the Order of the Bath (KCB). He died in 1794 at the age of 78, 2 years before his wife, Isobella. His son John studied medicine at the University of St. Andrews and succeeded him in his post at the Royal Haslar Hospital. The medical establishment continued to ignore Lind's results for 50 years, though experience had convinced many Navy officers and surgeons that citrus juices indeed protect from scurvy, even if the reason was unknown. It was to fall to another Scottish doctor, Gilbert Blane, to ensure that Lind's findings were put into practice.

Blane was born in Ayrshire and studied medicine under William Cullen at the University of Edinburgh, though he was awarded his MD, aged 29, from the University of Glasgow in 1778. He moved to London with a letter of introduction to the famous Dr. William Hunter, who helped Blane secure a post as private physician to Admiral Sir George Rodney. Blane served in Rodney's fleet with distinction in engagements in the Caribbean during the American War of Independence. During this period Blane, now promoted to Chief Physician for the West Indies Fleet, strove to improve the health of the sailors through improvements in nutrition and sanitation. He demanded monthly reports from all ships' surgeons and built a statistical analysis of the high levels of sickness that affected the fleet. In 1780 he compiled, published and distributed, at his own expense, a pamphlet on naval health and hygiene: *'On the most effective means for preserving the health of seamen, particularly in the Royal Navy'*. In it he wrote: "*Due attention to air, diet and cleanliness is not only more essential than mere medical treatment, but the sick cannot be considered as fit subjects for evincing the powers of medicine until they are provided for in this respect.*"

Despite the results of James Lind's clinical trial and his 1753 publication of '*A Treatise on the Scurvy*', the British Admiralty had ignored Lind's recommendations. Scurvy remained a major cause of sickness in the Royal Navy. Blane reported back in 1781 that one in seven seamen died from disease in the West Indies station in one year. His pamphlet, sent to all surgeons in the fleet, was vehement. Aware of Lind's findings, he summarised his recommendations succinctly: *"1st. The establishment of a certain method and discipline, in order to secure regularity and cleanliness among the men, and to render the ships clean and dry. 2ndly. The supply of fruit and other vegetables for the cure of the scurvy. 3rdly. The substitution of wine for rum. 4thly. The provision of an adequate quantity of necessaries for the sick. 5thly. The gratuitous supply of certain medicines. 6thly. The curing of certain diseases on board instead of sending them to hospitals; and Lastly. The preventing of filth, crowding, and the mixture of diseases in hospitals by proper regulations, and by establishing hospital ships."*

The pamphlet had at least some effect. A year after the recommendations were enacted in the West Indies, Blane could report to the Board of Admiralty that mortality levels were now reduced to one in twenty: *"the great squadron employed on this station has, by the attention of the Commissioners of Victualling, and also of the Commander in Chief, been supplied with most of the articles recommended in such quantities as to prove their efficacy; and indeed the small degree of mortality in comparison of former times is a sufficient demonstration of this."* Other officers of the fleet proposed to the Admiralty that Blane should receive special recompense and he was awarded a pension in recognition of his innovations in the West Indies fleet. Returning with accolades to Britain at the end of the War in 1783, Blane worked as a physician at St Thomas's Hospital in London and developed a large private practice. He became physician extraordinary to the Prince of Wales (and later to the King, George IV). He maintained his commitment to improving the health of seafarers. In 1785 in his paper *'Observations on the Diseases Incident to Seamen'* he wrote: *"But of all the articles, either in medicine or diet, for the cure of scurvy, lemons and oranges are much the greatest. They are real specifics in that disease, if anything deserves that name."*

In May 1794, the Royal Navy warship *HMS Suffolk* set sail with a small flotilla bound for India, fully supplied for daily rations of sugar mixed with lemon juice. The ships arrived in India after a four-month voyage without a single incident of scurvy. The news created a clamour.

With the British Navy's involvement in the French Revolutionary War requiring maximum capability and fitness, Navy fleet commanders demanded supplies sufficient for the daily issue of lemon juice to all naval seamen. Gilbert Blane was asked to join the Admiralty as a Commissioner on the 'Sick and Wounded Board', a post he held for 7 years. He quickly applied the lesson he had learned in the West Indies: the need for a scientific approach to observation and statistical reporting. He required every Navy surgeon to report the state of health of their ships by using standardised forms and medical logbooks. More than 600 surgeons served in the Royal Navy at that time, enabling Blane's pioneering use of statistics to be applied on an unprecedented scale. Data-gathering had become an officially mandated and centralised process.

Several years were needed to establish sufficient supplies of lemons and adequate methods of distribution to all ships in the fleet. By around 1800, systems were functioning, resulting in dramatic health improvements in sailors and gaining crucial advantage in naval battles against enemies who had yet to introduce similar initiatives. Lemons were augmented by limes from Britain's Caribbean colonies, (the origin of 'limey' as the common slang in America and South Africa, for the British), and for many years thereafter the preservation of the fruit juice was done by the addition of 15% of rum.

Blane resigned from the Navy for the second time in 1802, but continued to supply expert medical advice to the British government. He formulated regulations to control Plague in the Mediterranean, resulting in the evacuation of the British Army from Egypt, and in controlling 'Walcheren Fever', probably malaria, in the aborted British occupation of the Dutch island of that name during the Napoleonic Wars in 1809. He was knighted in 1812 and made baronet of Blanefield, his birthplace in Ayrshire. Near the end of his life he placed £300 in trust to establish the Gilbert Blane Medal for the *"advancement of medical science in the Royal Navy"* and judged the award of the first medals in 1832.

He died in London, aged 85, in 1834, having outlived his wife, Elizabeth, and seven of their nine children.

Despite the work of Lind and Blane which proved the 'antiscorbutic' protection provided by citrus fruit, scurvy continued to be a threat throughout the 19th century. Soldiers, seafarers and explorers continued to suffer wherever their leaders could not accept that the disease was fundamentally a nutritional deficiency. Vitamins were unknown (and not discovered until the 20th century) so the reason for the efficacy of citrus fruit was not understood. Furthermore, confusion was caused by the fact that other remedies were effective too. Napoleon's surgeon-in-chief reported that the consumption of horsemeat cured an epidemic of scurvy at the Siege of Alexandria in 1801. And taking the dried or distilled 'scurvygrass' plant *(cochlearia officinalis)* on board ships was partially effective; eating scurvygrass leaves or drinking its leaf tea became popular. We know now that *cochlearia officinalis* does indeed contain some of the crucial vitamin C (also known as ascorbic acid, precisely for the protection it provides against scurvy). Further confusion arose because, as it turns out, West Indies limes are lower in ascorbic acid than the lemons they replaced, and more vitamin C could be lost when their juices were exposed to air or copper pipes or vessels, rendering the juice less effective. Nevertheless in 1867, the British government passed the Merchant Shipping Act making it compulsory for all commercial British ships to carry lime juice. The same year, the third Scotsman in our story made an important innovation.

Lauchlan Rose was a chandler and merchant, born in Edinburgh's port of Leith in 1829 to a family of shipbuilders. He became one of a small number of merchants who supplied the Royal Navy with lime juice mixed with rum. In the year of the Merchant Shipping Act, he patented a process that allowed lime juice to be preserved without the addition of alcohol, but instead by the addition of sulphurous gas (sulphur dioxide) and sugar. This method not only improved the preservation of the juice, but opened up the emerging market for 'soft' drinks driven by the growing Temperance movement in Scotland and further afield. His first factory producing the juice was set up as 'L. Rose & Co.' in Leith the next year. The limes came to the port mainly from Dominica in the West Indies, and the success of "Rose's Lime Juice" was so great that in 1875 the company built new production and headquarter premises at Curtain Road in London, while still maintaining the Leith factory. In 1891 the Rose company purchased plantations in Dominica, (and then much later in West Africa) to ensure the adequate supply of limes. Rose's Lime Juice endures, as one of the products of the Coca-Cola Company.

Scurvy also endured despite the work of Lind, Blane and Rose. The confusion over the cause of the disease led to many more deaths, including that of William Stark, a Glasgow-educated doctor who had studied anatomy under John Hunter. In 1769, ignorant or dismissive of Lind's publication 16 years before, Stark embarked on a meticulously-documented programme of self-experimentation to determine the effect of diet on scurvy. None of Stark's 24 self-inflicted dietary regimes included fresh fruit or vegetables, and he died of scurvy and its complications eight months after the start of his experiments. It would not be until the 20th century that the cause of scurvy was finally determined. In 1927, the Hungarian biochemist Albert Szent-Györgyi isolated a compound, isolated from adrenal glands, that he called 'hexuronic acid'. He proposed this to be the crucial antiscorbutic agent, and received a Nobel Prize in 1937 after the connection between hexuronic acid and scurvy was confirmed by the American researcher Charles Glen King. Renamed 'scorbutic acid', we now know that scurvy can be cured with the daily dietary addition of just 10mg of that same essential ingredient - vitamin C.

13 Natural philosopher and natural educator: John Anderson (1726-1796)

Illustration 15: John Anderson Engraving by William Holl the Younger

John Anderson was a natural philosopher (today we would say physicist) and a liberalising educator who pioneered the application of science to technology in the early industrial revolution. He jointly founded the Royal Society of Edinburgh and was the posthumous founder of Anderson's University, which has evolved into the University of Strathclyde. Throughout his life he passionately promoted the education and advancement of working men and women. His dramatic public evening classes included many practical demonstrations, some of them literally explosive, and led to his nickname of 'Jolly Jack Phosphorus'.

Anderson was born in Dunbartonshire, the eldest child of the three sons and one daughter of James Anderson, minister of the parish of Rosneath, and his wife, Margaret Turner. His father died when John was about 8 years old and he was raised by an aunt in Stirling, where he went to the grammar school. He then attended the University of Glasgow, graduating in 1745. He returned to Stirling when the Jacobite uprising of 1745 led to trouble. When the army of Bonnie Prince Charlie besieged the town, in January 1746, the burgh council meekly surrendered, despite the fact that Stirling Castle was held by the

Hanoverian government forces. The government soldiers successfully defended the Castle, aided by a volunteer burgh corps of Hanoverians. The corps included the 20-year-old Anderson, who served as a gun-carrying officer. The Jacobite army retreated away to the Highlands and to its eventual destruction at Culloden.

Anderson seems to have returned for postgraduate study at Glasgow, before travelling to the Netherlands in 1750, then to London, where somewhat ironically he became private tutor to the 13-year-old Lord Doune, Francis Stuart, later the 9th Earl of Moray, who was a pupil at Harrow School. He spent three years at Harrow, lodging with his pupil in the house of the Master of the school, earning an annual salary of £40-£50 per year. As well as tutoring his pupil in classics, science and languages, he was responsible for managing the budget allowed by the 8th Earl, including Francis' pocket money of one guinea a week. This was a challenge. *"My Lord was almost every night at the play...or at Whist and was often a loser..."*. Next, Anderson was employed by Sir James Campbell as a travelling tutor for his son, with whom, as was common, Anderson went to France in 1754–5. There, Anderson learned that he had been appointed as Professor of Oriental Languages at Glasgow, mainly thanks to the patronage of the 3rd Duke of Argyll, Lord Ilay, who had been educated as a young man by Anderson's grandfather. It is not clear whether Anderson actively taught in this post (which was much later renamed the Chair of Hebrew and Semitic Languages), or whether this was an honorary position pending a more suitable appointment. In any event, in 1757 he transferred to the Chair of Natural Philosophy, reputedly causing controversy by voting for himself in the election. He was to hold the professorship for almost 40 years.

Anderson applied himself to his professorial duties with his usual energy and enthusiasm, and he was especially focused on engaging with local 'mechanics', and delivering public lectures. He encouraged a certain James Watt, who was then an instrument-maker at the University, and knew Benjamin Franklin. His usual class was highly mathematical, and he started a more practical, informal class for working people, which he called his 'anti-toga' class. These were extremely popular, and he delivered them in evenings twice a week, during academic sessions, for the rest of his life. It was said *"His style was easy and graceful, his command of language unlimited, and the skill and success with which his manifold experiments were performed, could not be surpassed. He excited the interest, and attracted the attention of his pupils, by the numerous and appropriate anecdotes with which he illustrated and enlivened his lectures. Enthusiastic in his profession, his whole ambition and happiness consisted in making himself useful to mankind, by the dissemination of useful knowledge..."* He was made a Fellow of the Royal Society in 1759, and published his *'Compend of Experimental Philosophy'* in 1760, followed by his textbook *'Institutes of Physics'* in 1777.

Throughout his tenure he involved himself vigorously in University of Glasgow administration and politics, where he could be argumentative and stubborn. He fought protracted battles with the Principal William Leechman, Professor of Divinity, a friend of Francis Hutcheson, and later Moderator of the Church of Scotland. Through the 1760s and 70s, Anderson pursued some of his fellow professors in the courts. The arguments raged over matters large and small. Anderson supported the French revolution, and that alone alienated some colleagues. He also viewed the management of University business and finance as appallingly deficient, and threatened to invoke a 'Royal Visitation' to examine the University's affairs. He was not alone in regarding its administration as lax, and its academic offerings old-fashioned. The 'College factor', Matthew Morthland, had been appointed in 1745 but never presented or settled the accounts on time. A group of professors began a case in the Court of Session to remedy the financial maladministration, and there was unhappiness among the students and alumni about the content and quality of teaching. One graduate of the University, the Reverend William Thom of Govan, produced a series of pamphlets attacking the University's curriculum and the professors' commitment to their students, writing *"A place [Chair] in a university is considered as easy, honourable and lucrative. It is almost looked upon as a sinecure: it is not ordinarily the most ingenious and able for teaching that [it] is pitched upon, but he who is connected...and can serve the men in power..."* Thom pointed to the example of Marischal College in Aberdeen, which he said benefited from the oversight of the Town Council, and the presence of a rival institution, King's College, in the same city. John Anderson came to agree with many of Thom's views and became increasingly argumentative in meetings of the Glasgow University faculty.

Examples of Anderson's disputes are in the University records and include: *'Statement by John Anderson against the regulations concerning degrees, 1 May 1764'; 'Minute of the protests lodged by John Anderson at the University of Glasgow meeting concerning the delays over disputes relating to the constitutional powers of Courts held by the Rector, the Dean and the Principal, 27 Feb 1766'; 'Memorial from John Anderson to the Principal and Professors of Glasgow College listing deficiencies in accommodation for the Physics class and for Physics instruments, 23 May 1768; 'Statement made at the Faculty meeting by Professor Anderson against the Principal's conduct in entering the minutes of the last meeting in a way contrary to statute and practice, 1778'; 'Copy letter from Principal Leechman, College of Glasgow, to Ilay Campbell, Lord Advocate, regretting the conduct of Professor John Anderson and assuring him that "with regard to a Royal Visitation, we are by no means apprehensive of the issue of any enquiry into our conduct and management", 1784;* and *'Copy letter from Principal Leechman, College of Glasgow, to the Marquis of Grahame seeking his co-operation against the evil actions of Professor John Anderson who has gone to London without asking permission of the Faculty, May 1785'.*

Anderson became embroiled in a number of angry feuds with fellow professors, and was impatient and intolerant of student unruliness or 'high jinks'. Nonetheless, he retained his aptitude and enthusiasm for teaching; but he became somewhat embittered and isolated from his colleagues. He had no family of his own, and when he drew up his will in 1795, he mandated that his estate be used to found an entirely new university in Glasgow, and specified in detail how his new institution was to be structured and managed. He intended his university to offer a modern and wide curriculum, including classes for women, and covering Medicine, Law, Theology and Arts – by which he meant Natural Philosophy, Mathematics, Chemistry and Logic as well as Greek and Latin. Disgusted by the politicking and mismanagement he had seen at the University of Glasgow, he mandated that no one associated in any way with that University was to serve in Anderson's University. He stated *"Thus...the almost constant intrigues, which prevail in the Faculty of Glasgow College about their Revenue, and the Nomination of Professors, or their Acts of Vanity, or Power, Inflamed by a Collegiate life, will be kept out of Anderson's University; and the irregularities, and neglect of duty in the Professors of Glasgow College, will naturally, in some degree, be corrected by a rival school of Education..."* He went even further, and forbade anyone involved in the governance of the City of Glasgow from holding office in Anderson's University – except those in the 'Tradesman' class. In a stipulation that would cause problems later, he also specified that his professors were not to be incorporated, so that they did not own the property of the new university nor could they be Trustees or managers of it, to avoid any conflicts of interest. The Trustees were not to permit professors *"as in some other Colleges, to be Drones or Triflers, Drunkards or negligent..."*

On Anderson's death in 1796 it was obvious that the full scope of his ambitious scheme was unaffordable, but his executors and 37 Trustees, including doctors, lawyers, merchants and tradesmen, nevertheless managed to establish the Andersonian Institution, as they decided to call it. The Institution began to offer courses in science in November of that year. Some of its early professorial appointments were inspired, and included Thomas Garnett, a superb lecturer and the first professor of natural philosophy, personally committed to the educational emancipation of women; his successor to that Chair, George Birkbeck, who continued his support for the adult education of men and women throughout his life, and founded the Mechanics Institute in Glasgow and Birkbeck College, in London; and Thomas Graham, the professor of chemistry who would inspire many of his students to follow his lead in the industrial applications of science. The

Andersonian added a medical school in 1800 (with David Livingstone among its alumni), that was later absorbed by the University of Glasgow. The Institution became Anderson's University in 1828, and grew successively into the Royal Technical College (1912), the Royal College of Science and Technology (1956) and the University of Strathclyde (1964), where John Anderson is remembered with respect and fondness in its eponymous physics building, in its Andersonian Library, and its main John Anderson campus near the centre of Glasgow.

14 The genesis of geology: James Hutton (1726-1797)

Since Christianity became Scotland's religion, it had been asserted that God had created the world in seven days. Accounts of Scottish history were referenced back to the Garden of Eden, and nobles, such as Sir Thomas Urquhart of Cromarty, claimed they could trace their ancestry back to Adam through 143 generations. In 1650, James Ussher, Archbishop of Armagh and Primate of All Ireland, reconstructed the history of the world based on the Bible, and pinpointed the creation of the Earth as occurring at 9 o'clock in the morning of Monday, October 23rd, 4004 B.C. Subsequently many Bibles had 'helpful' dates printed in scarlet in the margins, to remind readers of the supposed chronology of Biblical events.

Illustration 16: James Hutton by Abner Lowe

These beliefs were conventional until James Hutton, an Edinburgh-born farmer and chemist, examined some rocks on the south-east coast of Scotland. The chronology of the Earth, he realised, was not written in books, but in the rocks. Through a lifetime of observation and reasoning, he came to the conclusion that the Earth was continually being formed, over immense periods of time, through processes such as deposition, uplift, erosion and sedimentation. He pioneered the theory that became known as 'Uniformitarian' and is recognised as the 'father of modern geology'.

Hutton was born in Edinburgh into a prosperous family as one of five children to William Hutton, a merchant who was Edinburgh's city Treasurer, and Sarah Balfour. As with so many of his Enlightenment contemporaries, Hutton was just an infant when his father died, and he was raised by his mother, who encouraged her son's curiosity in nature and his abilities in chemistry and mathematics. He studied the classical and humanity courses in the University of Edinburgh from the age of 14. He had a brief spell as a lawyer's apprentice, which ended abruptly when he was found entertaining his fellow apprentices with chemical experiments. He became a physician's assistant, and then studied medicine at a leisurely pace in the universities of Edinburgh, Paris and Leiden, graduating, aged 23, in 1749. His many other interests intruded, and he never practiced medicine. He returned to Edinburgh and turned his knowledge of chemistry to practical advantage by starting a profitable business producing sal ammoniac (ammonium chloride) from soot, for use in the textile and farming industries. The business thrived, and together with his inherited wealth, made Hutton a rich man. But It was in fact mainly to gentlemanly farming that he devoted his next 20 years - somewhat by necessity, having inherited his father's farm in Berwickshire. His interest in agriculture and the land blossomed. He wrote that *"he was become very fond of studying the surface of the earth, and was looking with anxious curiosity into every pit or ditch or bed of a river that fell in his way..."* and he made several tours of Scotland, England and Flanders, studying mineral, fossil and rock formations as well as farming best practices.

Having made many improvements in the practice and management of his farm, he moved back to live in Edinburgh around 1767. He bought a share in the 'Company of Proprietors of the Forth and Clyde Navigation' and helped to direct that huge canal-building project, including sourcing the stone to build its 39 locks. He quickly became friendly with Edinburgh's luminaries of the Scottish Enlightenment. He co-founded the sociable and erudite 'Oyster Club' with Joseph Black and Adam Smith (who appointed Hutton and Black as his literary executors) and became friendly with James Lind, David Hume, the philosopher Adam Ferguson, and especially with the mathematician John Playfair, who would ensure Hutton's intellectual legacy. Perhaps influenced by Hume's rejection of miracles and the supernatural, Hutton was later to write: *"In interpreting nature, no powers are to be employed that are not natural in the globe...and no extraordinary events to be alleged in order to explain a common appearance."*

One prevailing theory of the Earth's formation, which purported to explain the discoveries of seashells and marine fossils in mountaintop rocks, was that all rocks had precipitated out of a single enormous flood - dubbed the 'Neptunist' theory. This was nicely in tune with the Old Testament stories of Noah's flood. An opposing view was held by the so-called 'Plutonists' who pointed out that earthquakes and volcanoes continually reshaped the earth; and anyway, what had happened to all the water that had supposedly flooded the mountain ranges? Hutton observed evidence that would settle the debate - or should have - and made some exceptionally acute deductions. At Glen Tilt in the north of Perthshire, and in the Salisbury Crags near his home in Edinburgh, Hutton found veins of granite penetrating deep into metamorphic marble 'schists'. This showed that the granite had been fluid, formed from the cooling of molten rock, rather than precipitating from water, and that the granite must be younger than the schists. He recognised that the interior of the Earth must be hot, and that this heat must drive the creation of new rock, which was then eroded by air and water and deposited as sediments in the sea. The Earth's internal heat then consolidated the sedimentary layers into stone on the seabeds, and somehow uplifted it to form new land and even mountain ranges, containing the evidence of their aquatic history.

Hutton presented these ideas in a paper of two parts called *'Theory of the Earth'* to the Royal Society of Edinburgh in 1785. The first part was presented by his good friend Joseph Black, the second part by himself.

A month later Hutton summarised his theory in a shorter paper *'Concerning the System of the Earth, its Duration and Stability'*. He wrote: *"The solid parts of the present land appear in general, to have been composed of the productions of the sea, and of other materials similar to those now found upon the shores. Hence we find reason to conclude: 1st, That the land on which we rest is not simple and original, but that it is a composition, and had been formed by the operation of second causes. 2nd, That before the present land was made, there had subsisted a world composed of sea and land, in which were tides and currents, with such operations at the bottom of the sea as now take place...Lastly, that while the present land was forming at the bottom of the ocean, the former land maintained plants and animals; at least the sea was then inhabited by animals, in a similar manner as it is at present. Hence we are led to conclude, that the greater part of our land, if not the whole had been produced by operations natural to this globe..."*

The implications for the age of the Earth were clear, though Hutton's prose was often tortured and obscure to say the least. The Earth had to be at least millions of years old, not the 6,000 years of Biblical mythology. We now know, of course, the true age of the Earth's oldest rocks is more than 4,000 million years.

Hutton's major work *'Theory of the Earth'* was not published until 1795, and such was the obscurity of the language he used, that its arguments were difficult to fathom. It was for another of his friends John Playfair, professor of mathematics at the University of Edinburgh, to explain and clarify Hutton's conclusions in his *'Illustrations of the Huttonian Theory of the Earth'*, published in 1802. Playfair introduced his work with the words, *"The Treatise here offered to the public was drawn up with a view of explaining Dr. Hutton's 'Theory of the Earth' in a manner more popular and perspicacious than is done in his own writings. The obscurity of these has often been complained of... so little attention has been paid to the ingenious and original speculations which they contain."* Hutton had died 5 years before, but he had gathered more evidence to support his theories, by discovering rock 'unconformities' including on the island of Arran, at Inchbonny near Jedburgh and (with Playfair) at Siccar Point near Dunbar. These unconformities displayed diverse types of rock, volcanic and sedimentary, overlaid on each other at various angles of tilt. Hutton's remarkable insight was that the processes that had created the Earth were hugely ancient, and that they had *"No vestige of a beginning, no prospect of an end."* This view became known as 'uniformitarian', an ugly word to describe the view that geological processes were continuous, ongoing, usually gradual, and happened over immense spans of time.

Hutton's curiosity and perceptiveness extended beyond geology. He studied atmospheric changes, explaining the formation of rain, and hypothesised that biological and geological processes are interlinked. He was said to have viewed the Earth as a kind of connected organism, presaging the Gaia theory of James Lovelock. In his later writings he proposed that the development of life on Earth was as progressive and continuous a process as the formation of rocks: *"...if an organised body is not in the situation and circumstances best adapted to its sustenance and propagation, then, in conceiving an indefinite variety among the individuals of that species, we must be assured, that, on the one hand, those which depart most from the best adapted constitution, will be the most liable to perish, while, on the other hand, those organised bodies, which most approach to the best constitution for the present circumstances, will be best adapted to continue, in preserving themselves and multiplying the individuals of their race."* This idea was to influence two more Edinburgh-trained physicians, named Erasmus and Charles Darwin.

James Hutton never married, but had a son, James Smeaton Hutton, with a Miss Edington, around 1747. Hutton provided financial assistance but did not help to raise his son, preferring to live with his four sisters throughout most of his time in Edinburgh. He died without a will after an illness of many years, which had been lessened, but not cured, by a bladder operation done by Joseph Black. Hutton passed on his significant wealth and property to his only surviving sister Isabella. He is revered as the father of the modern science of geology, and gives his name to the James Hutton Institute, a globally-recognised research organisation promoting the sustainable use of land and natural resources. His theory of the earth was resisted for many years by natural philosophers who clung to the view that the formation of the Earth's surface was due to a series of sudden 'catastrophes', and his theory offended many Christians by rejecting the biblical account of creation and its supposed subsequent chronology. However Hutton's pioneering work ignited widespread interest in this new science of geology, and his ideas would be taken up by the next generation of geologists - notably by another far-sighted and perceptive Scotsman - Sir Charles Lyell.

15 Shedding light on heat, and the 'discovery' of air: Joseph Black and Daniel Rutherford (1729-1819)

Joseph Black was born in Bordeaux, France, the sixth of 13 children of Margaret Gordon from Aberdeenshire and John Black, a wine merchant of Scots descent from Belfast. The family were comfortable and well-connected in France, and Joseph said of his parents: *"My father...had no ambition to be very rich; but was cheerful and contented, benevolent and liberal-minded. He was industrious and prudent in business, of the strictest probity and honour, very temperate and regular in his manner of life. He and my mother, who was equally domestic, educated thirteen of their children, eight sons and five daughters, who all grew up to men and women, and were settled in different places. My mother taught her children to read English, there being no school for that purpose at Bourdeaux."* Joseph was educated at home until he was 12, and then entered grammar school in Belfast. At the age of 18 he attended the University of Glasgow for four years and was influenced by the great teaching of William Cullen, physician and newly-appointed lecturer in chemistry, who became a life-long friend and colleague. In 1751 Cullen became Regius Professor in the Practice of Medicine at Glasgow, but Black went on to study medicine at the

Illustration 17: Plaque of Joseph Black by James Tassie (Hunterian Museum, Glasgow)

University of Edinburgh, graduating with a ground-breaking doctoral thesis on the use of the salt 'magnesia alba' (now known to be magnesium carbonate) in the treatment of kidney stones.

A year later after some more experiments, Black's work was presented before the Medical Society of Edinburgh and then published in the second volume of his *Essays and Observations* as *Experiments upon Magnesia alba, Quicklime, and some other Alkaline Substances*. To appreciate the significance of Black's results we should understand the state of chemical knowledge at the time. Matter was believed to consist of five principal elements: Water, Salt, Earth, Metal and Fire. All combustible matter was thought to contain a substance called 'phlogiston' which was released when matter burned. Some living things, such as growing plants, were supposed to absorb phlogiston, which was said to be why air does not spontaneously combust and also why plant matter can be burned.

Similarly the caustic nature of alkaline substances was believed to be due to their phlogiston content. For example quicklime (made through the heating and decomposition of limestone) was regarded as chalk which had taken up phlogiston; and when mild alkalis such as soda ash or potash were made more caustic by mixing with quicklime, the phlogiston was supposed to pass from it to them. As a student Black had developed sensitive analytical balances to weigh chemical reactants and products. His alkali experiments disproved accepted belief. When heating magnesia alba to produce caustic quicklime, measurements proved that weight was lost. Something had gone missing, which he found to be an 'air,' which, because it was fixed in the magnesia before it was 'causticised', he described as 'fixed air'. He further showed that the lost weight was restored when the quicklime was made to reabsorb the fixed air it had released. In another classic experiment, he demonstrated that one cubic inch of marble decomposed into half its weight of pure lime and a volume of 6 gallons of 'fixed air'. His tests showed that the 'fixed air' was denser than normal air and supported neither flames nor animal life. When bubbled through limewater it would precipitate chalk. He used that fact to show that 'fixed air' is produced by animal respiration and microbial fermentation, and referred to it also as 'mephitic air' meaning noxious. He had in fact for the first time isolated the gas we know as carbon dioxide. The experiments also showed that magnesium and calcium were different substances, contrary to accepted opinion. Black wrote: *"We have already shewn by experiment, that magnesia alba [magnesium carbonate] is a compound of a peculiar earth [magnesium] and fixed air [carbon dioxide]"*.

These findings made Black's reputation as a great experimental scientist. In 1756 he was elected to the professorial Chair of Chemistry at Glasgow, just vacated by William Cullen, who had moved in the other direction to the Chair of Chemistry and Medicine at Edinburgh. News spread fast of his discovery of 'fixed air', by now being also called 'carbonic acid gas'. The French chemist Antoine Lavoisier, who would go on to isolate oxygen and hydrogen, which finally destroyed the phlogiston theory, wrote to Black to report some of his own experimental findings on respiration, saying *"It is but just you should be one of the first to receive information of the progress made in a career which you yourself have opened, and in which all of us here consider ourselves your disciples"*. In 1757, Black was further honoured by the University of Glasgow by being appointed the Regius Professor of the Practice of Medicine.

More important findings were to be made. Black applied heat to ice at its melting point and measured no rise in the temperature of the ice/water mixture, but observed that the heat just caused the production of more melt-water. Similarly, Black showed that heating boiling water does not result in a rise in temperature of the water/steam mixture, just an increase in the amount of steam. Where did the applied heat go? Clearly the heat must have been absorbed by the ice particles and boiling water and become 'hidden' or as he termed it, 'latent'. Conversely, when steam condenses on a colder surface, its latent heat is transferred to the surface material. Black also noticed that different substances of equal mass need different amounts of heat to raise their temperature by the same degree - the concept of specific heat capacity. This and the discovery of latent heat prompted the first stirrings of the science of thermodynamics.

The theory of latent heat proved to be of great practical importance too. In around 1757, Black was introduced to a young man called James Watt, who was an instrument-maker at Glasgow University. The latent heat of water is much larger than most other liquids, and Black was to provide both technical consultancy and financial support to Watt's ingenious efforts to improve the efficiency of steam engines, as we shall see.

Black once more followed his friend and mentor, William Cullen, when in 1766 he succeeded Cullen to the Chair of Medicine and Chemistry at the University of Edinburgh. Focusing now more on teaching and his select private medical practice, rather than research, Black was soon gathering large and enthusiastic audiences of students. His Edinburgh colleague John Robison, Professor of Natural Philosophy, wrote of Black's success as a teacher: *"It could not be otherwise...His personal appearance and manner were those of a gentleman, and peculiarly pleasing. His voice in lecturing was low, but fine; and his articulation so distinct that he was perfectly well heard by an audience consisting of several hundreds. His discourse was so plain and perspicuous, his illustrations by experiment so apposite, that his sentiments on any subject never could be mistaken, even by the most illiterate; and his instructions were so clear of all hypothesis or conjecture, that the hearer rested on his conclusions with a confidence scarcely exceeded in matters of his own experience."* As Black advanced in years, Robison noted, he *"preserved a pleasing air of inward contentment...he was of most, easy approach, affable, and readily entered into conversation, whether serious or trivial."* Black was fully engaged in the Edinburgh societies of Enlightenment thinkers, being particularly friendly with his friend and former colleague at Glasgow, Adam Smith. He and Smith were members of the Oyster Club and the Poker Club (named after the fireplace implement not the card game) together with such as Adam Ferguson, David Hume, John Robison, and James Hutton. The Poker Club had previously been the Militia Club, founded to promote the formation of a Scottish Militia (on the model of those in England) to protect against further Jacobite insurrections and possible Civil War. This never happened and in its renamed form the Poker Club seems to have been a benignly relaxed dining and drinking establishment where *"... The dinner was set soon after two o'clock, at one shilling a head, the wine to be confined to sherry and claret, and the reckoning to be called at six o'clock."*

One of Joseph Black's most promising students at the University of Edinburgh was Daniel Rutherford, the son of Anne Mackay and John Rutherford, Black's professorial colleague at the Edinburgh Medical School. Young Rutherford followed his father's calling and studied medicine under Black and William Cullen, graduating aged 23 in 1772. When a final-year student, he was asked by Black to continue the investigations into the 'fixed air' that Black had isolated 18 years before. Rutherford embarked on an experiment to remove from normal air both the life-supporting oxygen and Black's 'fixed' or 'mephitic' air we know as carbon dioxide. Rutherford held a mouse in a closed container until it died. He burned a candle and phosphorous in the remaining 'mephitic air' and passed it through 'lye', a caustic solution that was known to absorb fixed air. When this was done, however, there was still a significant amount of gas remaining, which, he showed, extinguished flames and did not support the life of another mouse. He had isolated a component of air that was not Black's carbon dioxide, but something different. Rutherford called the gas 'noxious air', and had in fact had isolated nitrogen. He followed the accepted wisdom of the times and assumed that the noxious air was saturated with phlogiston, the product of burning and respiration. Despite this incorrect assumption, he is credited with the discovery of nitrogen, which makes up around 80% of the earth's atmosphere. His doctoral dissertation in 1772 was on the subject of his experiments: *'De Aere Fixo Dicto, aut Mephitico'*.

Rutherford went on to a distinguished career in his own right, becoming a Fellow of the Royal College of Physicians of Edinburgh in 1777 and then becoming Professor of Medicine and Botany and keeper of the Royal Botanic Garden in Edinburgh in 1786, (even though his interest in botany was said to be limited and his botany courses were criticised and poorly attended). In 1786 he was married to Harriet Mitchison, and later became the first Professor in the Practice of Physic at the University of Edinburgh. He died suddenly in 1819, aged 70.

Joseph Black never married, and died peacefully at home in Edinburgh in 1799, aged 71. In later life he was a co-founder of the Royal Society of Edinburgh, President of the Royal College of Physicians of Edinburgh, and principal physician in Scotland to King George III. His friend Adam Smith said of him *"no man had less nonsense in his head than Dr. Black."* He is remembered in the names of the chemistry department buildings at both the universities of Glasgow and Edinburgh.

16 Steam power efficiency: James Watt (1736-1819)

The popular story goes that James Watt observed the rattling lid of a boiling kettle, and impressed by the power of the steam, was inspired to invent the steam engine. This is a nice story, but it is a myth. The first working steam engine was developed in 1712 by an Englishman, Thomas Newcomen, who devised a steam-powered water pump building on the earlier inventions of Denis Papin and Thomas Savery. James Watt's genius was in recognising the inherent limitations of Newcomen's design, and in using his knowledge and skill to develop an entirely new type of steam engine, one which was destined to become the driving force behind the Industrial Revolution.

Illustration 18: James Watt by John Partridge

Watt grew up in Greenock, where his family was prominent. His father, also James, was a shipwright, chandler, and served as a local baillie (that is, a magistrate). His grandfather, Thomas Watt, was well-respected in Greenock, a teacher of mathematics, surveying, and navigation. As a young man Thomas had been dispossessed of his Aberdeenshire farm because his own father had fought (and died) for James Graham, the Royalist Marquis of Montrose, in his battles against the Covenanters in 1644-45. James was the eldest surviving child of the eight children of Agnes Muirhead and James Watt Senior, who lost five children in infancy. James was frail, frequently ill and initially taught at home by his mother, who had a strong and intelligent personality, before going to school where he was regarded as 'slow', especially in Latin and Greek. But he shared his grandfather's strong aptitude in mathematics, and on leaving school showed practical mechanical skills while working in his father's chandlery, becoming familiar with its nautical paraphernalia of pulleys, compasses, sextants and quadrants.

In 1755 Watt's mother died at the age of 52, and his father's health was ailing. The business was ailing too, and Watt needed to find a paying job. It was decided that he should go to Glasgow and become an instrument-maker. Contacts and nepotism always help, and George Muirhead, the Professor of Humanity at the University of Glasgow, was related to Watt's mother. Muirhead introduced Watt to his fellow professors, John Anderson and Robert Dick, the professor of natural philosophy. Professor Dick immediately saw the potential in the young Watt and advised him to travel further afield to train in instrument-making.

Watt, aged 19, arrived in London in 1755 and having never completed an apprenticeship, he initially struggled to find a position. He was in the end introduced to John Morgan, a mathematical instrument maker, who agreed to instruct Watt for a year in return for Watt's labour plus a fee of 20 guineas. Thus began a period of intensive, uncomfortable and cash-strapped training. In his time in London, Watt seems to have kept his head down. Apart from his focus on work, there were good reasons to maintain a low profile. Military tensions were building, and Britain, with Prussia and Hanover, declared war on France and her allies in May of 1756, the beginning of the Seven Years War. Naval press gangs were active in the city, and between 1755 and 1756 more than 70,000 men were recruited, around half through press-ganging. Although even in wartime, tradesmen and legitimate apprentices stood a good chance of avoiding impressment, Watt was not officially an apprentice and was vulnerable. His unofficial status, if discovered, would have caused problems too with the London trades guilds.

Nevertheless by the end of his year in London, Watt wrote: *"I am now able to work as well as most journeymen"*. He returned to Glasgow in 1756, unwell, impoverished, and jobless. Again Professor Dick stepped in, and he helped Watt to open a workshop in the University, and to become its mathematical instrument-maker. Dick provided the first task, to restore an astronomical clock that was in poor condition, and introduced James to more of the various eminent Glasgow professors of the day: Adam Smith, the moral philosopher and economist; Joseph Black, professor of chemistry and medicine; Robert Simson, professor of mathematics; and a precocious young student, John Robison, who would succeed Black as professor of chemistry and later became professor of natural philosophy at the University of Edinburgh. On his meeting with Watt in his workshop, Robison later wrote: *"I saw a workman, and expected no more; but was surprised to find a philosopher, as young as myself, and always ready to instruct me. I had the vanity to think myself a pretty good proficient in my favourite study, and was mortified at finding Mr. Watt so much my superior... Whenever any puzzle came in the way of the young students, we went to Mr. Watt...and we knew he would not quit it till he had either discovered its insignificancy, or made something of it. He learnt the German language to peruse Leopold's 'Theatrum machinarium'. So did I, to know what he was about. Similar reasons made us both learn the Italian language..."*.

Watt's obvious talents and diligence led to plenty of work from the Glasgow professors, particularly Joseph Black and John 'Jolly Jack Phosphorus' Anderson who succeeded Dick as professor of natural philosophy, and always required equipment to illuminate his lectures with practical demonstrations. Black, for his part, would become a life-long friend and financial investor in Watt's later enterprises. But it was John Robison who would draw Watt's attention to the potential power of steam engines. Robison, as what we would now call a postgraduate student, had published notes on an improved design of Newcomen's rudimentary steam-powered water pump, and had the idea of using steam power to make a motor vehicle. Watt was much later to write: *"My attention was first directed, in the year 1759, to the subject of steam-engines, by the late Dr. Robison, then a student in the University of Glasgow, and nearly of my own age. He at that time threw out an idea of applying the power of the steam-engine to the moving of wheel-carriages, and to other purposes; but the scheme was not matured, and was soon abandoned on his going abroad...".* But a seed of interest was planted in Watt's mind and he began experimenting over the next few years.

In 1763 Watt was asked to repair a model of a Newcomen engine that was owned by the University. Newcomen's so-called 'atmospheric' engines consisted of a single vertical cylinder enclosing a piston which was balanced to sit at the top of the cylinder. The piston was sucked down by injecting low-pressure steam and then condensing it with a spray of cold water, creating a partial vacuum below the piston. The combination of vacuum below and atmospheric pressure above drove the piston down, before it returned to the top of the cylinder for the next 'power' stroke. The piston was attached via a rocking bar at the top of the engine, to a vertically-driven water pump, commonly applied to keep mine-workings dry. The full-sized engines were temperamental, difficult to start, inefficient in their use of steam, and hence wasteful of the coal fuel needed for the boiler. The small model in the possession of Glasgow University also had problems. John Anderson asked Watt to have a look at it, and make the model work well.

Watt quickly established that it was difficult to keep even the small model (cylinder size 6x2 inches) adequately supplied with steam. He patched the model up, and continued to think and experiment. He rightly concluded that in Newcomen engines much steam was wasted in warming the cylinder, which had just been cooled by the spray of condensing water. By repeatedly heating and cooling the cylinder, the engine wasted most of its steam rather than producing the required vacuum. Watt imagined the construction of a 'perfect engine': one that consumed just a single cylinder of steam to produce a perfect vacuum. He made a new model, larger than the original, and conducted experiments on it to study the mechanisms of steam production and condensation, sometimes indeed using a tea kettle as a boiler. He discovered the practical implications of the high latent heat of steam and the heat capacities of the cylinder material. For example, the condensing steam raised the temperature of the injected condensing water, and this combined with the heat stored in the cylinder meant far more condensing water was needed than expected, and far more steam was required in the next cycle. While not initially understanding the underlying scientific principles of latent heat and the specific heat capacity of materials, he soon learned them from Joseph Black.

The engineering solution came to him during a Sunday afternoon stroll on Glasgow Green in the spring of 1765 - namely, an entirely separate condenser that would avoid the inefficient heating and cooling of the cylinder on each cycle of the engine. Around forty years later, he would say, that once the separate condenser was conceived, *"all ... improvements followed as corollaries in quick succession, so that in the course of one or two days the invention was thus far complete in my mind, and I immediately set about an experiment to verify it practically."* The improvements included a 'steam jacket' around the cylinder to maintain it at the same temperature as the injected steam. In April he wrote to his friend Dr. James Lind, (who was then already working in Gosport Hospital) *"I have now almost a certainty of the facturum of the fire-engine, having determined the following particulars: the quantity of steam produced; the ultimatum of the lever engine;...the quantity of steam destroyed by the cold of its cylinder; the quantity destroyed in mine... in short, I can think of nothing else but this machine".* By September, Watt told Lind about his experiments with a new model of his 'perfect' engine, saying that he had *"...tried..a small model of my perfect engine... almost totally to prevent waste of steam and consequently to bring the machine to its ultimatum."* But many practical problems needed to be solved if full-scale working engines were to be developed.

Watt needed financial backing to fund more development. His instrument-making business had prospered - by 1764 he was employing 16 men in his Glasgow workshops, producing and repairing quadrants, compasses and musical instruments. His business partner was a merchant, John Craig, who had made loans, to be repaid with interest, of £100 annually over several years to fund the expansion of the business. This stability enabled Watt to marry his cousin Margaret Miller (known as Peggy) in 1764; they would have a daughter, also Margaret, and a son, James Jr. By the time of his marriage, Watt was also a fully-fledged participant in the Enlightenment gatherings of Scottish intellectuals, such as the Glasgow Literary Society, whose illustrious members included Adam Smith, David Hume, John Robison, John Anderson and Joseph Black, and he became friendly with the geologist James Hutton. Here was the opportunity to network and find the new backers he needed, a need made more urgent by John Craig's untimely death in December 1765. Craig's trustees demanded the repayment of his loans, totalling £757, and further development of the 'perfect' engine would require far more.

Joseph Black had been another early investor in Watt's ideas, but had lost a small fortune in the European banking crisis that followed the end of the Seven Years War in 1763. Black introduced Watt to one of his former pupils, John Roebuck, a successful businessman. Roebuck was founder and owner of the Carron Ironworks near Falkirk, and had a need: his coal-mines at Kinneil near Bo'ness were flooded, and he did not think Newcomen engines would be up to the job of pumping them out. He offered Watt the chance to develop a working engine at Carron and at Roebuck's own home, and adopted Watt's financial debts to John Craig's trustees. Further, more serious, funding would be contingent on Watt's progress.

Progress was initially slow. To develop a commercially viable full-scale engine, Watt needed not only funds, but also practical experience in civil engineering. So as well as the experiments at Carron, Watt involved himself in setting up and modifying the first Newcomen engines to be installed in Scotland, at Falkirk and Ayr, and in consulting on the boom industry of the day, canal construction. He attended Parliament in connection with the Forth and Clyde Canal Act of 1768 and wrote to his wife Peggy: *"I think I shall not long to have anything to do with the House of Commons again—I never saw so many wrong-headed people on all sides gathered together."* Of his canal work Watt said: *"I would not have meddled with it had I been certain of being able to bring the engine to bear; but I cannot, on an uncertainty, refuse every piece of business that offers."* Nevertheless his experiments and the construction of a new model test engine persuaded Roebuck to pay for a patent application in return for two-thirds of the intellectual property and future profits arising. The patent of 1769, entitled *'A New Method of Lessening the Consumption of Steam and Fuel in Fire Engines'*, was controversial because it was drawn so widely. Rather than specifying a particular design, it claimed coverage of practically all steam engines using a separate condenser, and would provoke legal disputes in the years to come.

Business trips to England to study innovations and support the patent application resulted in the most significant meetings of Watt's career. He met Roebuck's friend, the polymath medic William Small, in Birmingham. Small had been introduced, by Benjamin Franklin, to a Birmingham entrepreneur, Matthew Boulton, and Small returned the favour to Watt. Boulton's huge 'Soho' factory in Birmingham was a marvel of organisation, producing clocks, coins, buttons and ornaments, and employing 600 people. Boulton was an enthusiastic innovator and saw the potential to replace human labour with automation, for which his factory needed power. By the spring of 1770 Watt's engine at Roebuck's Kinneil House was working well with a separate condenser. The details were shared in strict confidence with Boulton and Small, who proposed that they should join the established partnership of Watt and Roebuck. Negotiations were protracted due to Roebuck's resistance, and Watt wrote: *"Nothing is more contrary to my disposition than hustling and bargaining with mankind, yet that is the life I now constantly lead."* Over the next years Boulton followed Watt's progress intently.

The year 1773 was to prove the most traumatic of Watt's life. In March, Roebuck's businesses ran into near-bankruptcy, which forced him to deal with Boulton, who acquired his two-thirds share of the steam engine patent. The Kinneil engine was dismantled, packed up and dispatched to Boulton's Birmingham factory. Much worse, in September Watt's wife Peggy died in childbirth and the child was stillborn. Watt, shaken and depressed, decided to accept Boulton's offer to re-start work on the engine, but in Birmingham. The next spring, Watt travelled to England with his old friend, James Hutton, as a travelling companion. In October his daughter Peggy, aged eight, and his son, James Jr. aged six, joined him in Birmingham and James Jr. was sent to a school in Winson Green as a boarder.

The steam engine patent was already five years old with eight years of protection remaining. Boulton wanted an extension and suggested it was better and cheaper to seek an Act of Parliament to extend the patent (at a cost £110) rather than registering a new one (cost £130). With the help of Boulton's connections in Parliament, an Act was passed in May 1775, against much opposition from rivals and vested interests, and the patent was extended for a further 25 years. A month later, the business partnership of Boulton and Watt was formed. Watt's morale had recovered and he entered into Birmingham's social life, in particular the 'Lunar Society' of industrialists and scientists, which met for dinner every month on the Monday closest to the full moon, and included Boulton, Small, Erasmus Darwin, Josiah Wedgewood and (occasionally) James Hutton and Benjamin Franklin. In 1776, Watt married Ann Mcgregor, with whom he would have seven children. At that time too, he recorded the terms of his partnership with Boulton. Boulton was to adopt the financial risk for the expense of the 1775 Act of Parliament and all the costs of future research, and be responsible for the business management and accounting. Watt was in charge of design, engineering direction, surveys and installations.

The 'Boulton & Watt' company embarked on a rapid programme of development, helped by the recruitment of a team of "Engineers" to assemble, install and tune the engines produced by the factory - arguably marking the birth of engineering as a profession. Foremost among the recruits was a talented young Scottish inventor, William Murdoch, born near Cumnock, Ayrshire, in 1754. At the age of 23, Murdoch walked to Birmingham, a distance of more than 300 miles, to ask Watt for a job. His intellect and energy were impressive (Boulton was said to have been impressed by Murdoch's wooden hat, made on a lathe of his own design!) and he was hired. Murdoch's recruitment was a crucially important factor in the success of Boulton & Watt. Initially working in the pattern shop, Murdoch quickly became indispensable in managing engine installations (his first one at Wanlockhead lead mine in Dumfriesshire) and in suggesting many design improvements to gears and valves. The company's key target market was the tin-mining industry of Cornwall, where many Newcomen steam-driven water pumps were installed. Murdoch, Boulton and Watt made many sales visits there, so many that Watt and Murdoch set up Cornish bases. Watt rented Cusgarne House near Truro and Murdoch took a house in Redruth (which was the first domestic house to be lit by coal gas, which he produced by heating coal and piping the gas through an old gun barrel). In Redruth, Murdoch also worked in his spare time on developing a steam-powered carriage. Witnesses in 1784 saw a model 3-wheel steam carriage run around Murdoch's living room - the first recorded example in Britain of a self-powered moving vehicle.

Watt was uncomfortable as a salesman and the cost-model to set the price of the engines was innovative but complex. The efficient Boulton & Watt engines consumed less fuel than the Newcomen pumps, so the price of the royalty to use the Watt engine was set at one-third of the cost of the fuel saved. Much early business was based on making improvements to old Newcomen pumps, for little profit. It took nearly 20 years before Boulton & Watt could manufacture and supply complete engines from the Birmingham factory. During that time many more engineering innovations were made, including the adaption to produce rotary motion using 'sun and planet' gears, an idea patented by Watt but conceived by Murdoch.

Commercial success came slowly but surely, involving more engineering innovations and patent battles. The market widened from the focus on mines to the need for mechanisation in factories and mills. Boulton and Watt established their own 'Albion' Flour Mill in London, with two rotary Watt engines powering milling machinery designed by yet another Scottish engineer, John Rennie, who was a pioneer in substituting cast iron for wood in complex structures. The use of efficient steam engines in place of unreliable water-mills and horse-driven power required the standardisation of the concept of power itself. Watt estimated that an 'average horse' could work sustainably raising 22,000 pounds of weight by 1 foot in 1 minute. Adding 50% for good measure, to ensure that he was not accused of exaggerating the capability of the new engines, he fixed the standard 'horsepower' at 33,000 foot-pounds per minute. The horsepower of each Watt engine was then established and a premium of £5 per horsepower per year was charged for engines covered by the 1769 patent. Today of course the international standard (SI) unit of power is the Watt, (1 Joule per second) - equivalent to around 746 horsepower in Watt's original definition.

Watt went into semi-retirement in 1800, the year that the original steam engine patent expired and his partnership with Boulton ended. The partnership was carried on successfully by their sons, James Jr. and Matthew Jr., with William Murdoch (now Murdock) made a partner. By then the Industrial Revolution, powered by steam, was in full swing. Cotton spinning became a major consumer of steam power - by 1818 Glasgow alone had 18 weaving factories containing 2,800 looms. In retirement Watt continued his wide interests in the Lunar Society, and grew his interest in medical science, reinforced by the deaths from consumption (tuberculosis) of his daughter Jessie at the age of 15 in 1794, and his son Gregory, a promising geologist, in 1804, aged 27. He travelled in Europe with his wife, and visited Scotland frequently. Now a very wealthy man, he had acquired Heathfield Estate in Staffordshire and died at home in Heathfield Hall in 1819. Many of his friends had predeceased him: Hutton in 1797, Black in 1799, Robison in 1805 and Boulton in 1809, at the age of 81.

Watt died widely recognised and greatly honoured, a Fellow of the Royal Society of Edinburgh, an honorary Doctor of Laws at the University of Glasgow, and a Foreign Associate of the French Academy. As a founding father of the Industrial Revolution and the profession of engineering, his name is given to the titles of many engineering institutes worldwide, including that of Heriot-Watt University in Edinburgh, which now houses his memorial statue originally placed in Westminster Abbey. The eulogy in the Abbey is dedicated to *"James Watt who, directing the force of an original Genius, early exercised in philosophic research, to the improvement of the Steam Engine, enlarged the resources of his Country, increased the power of Man, and rose to an eminent place among the most illustrious followers of science and the real benefactors of the World."*

17 Cerebral geniuses: Alexander Monro *secundus* and George Kellie (1733-1829)

Illustration 19:Alexander Monro (secundus) - engraving by James Heath after Henry Raeburn

Alexander Monro is one of the most famous names in the history of medicine. To be exact, it is three of the most famous names. The Monro dynasty was to hold a Chair of Medicine at the University of Edinburgh for a total of 126 years. The first, known as Alexander Monroe primus, studied medicine in Edinburgh, London, Paris and Leiden, and became the founding Professor of Anatomy at the new Edinburgh Medical School in 1720. His father, John Monro, was a military surgeon in the armies of William III and George I. He was wounded in the Battle of Sheriffmuir during the 1715 Jacobite rebellion, and later was instrumental in the creation in Edinburgh of a 'Seminary of Medical Education' modelled on the medical school of the University of Leiden, where he had trained. The appointment of his son, Alexander primus, to the first chair of anatomy at the Edinburgh Medical School was the culmination of long-term political manoeuvring by John, and extended studies and apprenticeships by his son. The professorship of Alexander Monro primus proved highly successful. He lectured in English rather than Latin, and the popularity of his lectures and demonstrations increased the demand for cadavers for dissection. This prompted an outbreak of grave-robbing and, in turn, public demonstrations against dissection, forcing Monro primus to teach solely within the safe confines of Edinburgh University. In his distinguished career he published a seminal textbook *The Anatomy of the Human Bones*, and a notable paper *'An Account of the Inoculation of Smallpox in Scotland'* where he laid out the case for inoculation, controversial at the time. He was elected a Fellow of the Royal Society, and established a small teaching hospital which became the Royal Infirmary of Edinburgh. Its first patients included the wounded from the Battle of Prestonpans in 1745, where the Jacobites of Charles Edward Stuart won a quick victory over Government troops led by Sir John Cope. Monro primus, like his father, John, was a staunch supporter of the Hanoverian government; but the casualties from both armies were treated, both in the battlefield and in the new Royal Infirmary.

In 1725, aged 28, Monro had married Isabella MacDonald, three years his senior and the third daughter of Sir Donald MacDonald of Sleat. The marriage produced three sons and five daughters, though only a single daughter survived to adulthood. Their youngest son, Alexander *secundus,* born in 1733, and the man who would be one of his accomplished students, George Kellie, (born in 1770) are the main subjects of this story.

Alexander junior (and his brothers) were precocious students and his father's immediate ambition was for Alexander to succeed him in his medical professorship at Edinburgh. The boy attended the University at the age of 12, studying the general 'philosophy' course including mathematics under Colin Maclaurin. By age 18 he was assisting his father in human dissections. The popularity of his father's lectures had created huge demand, and a need to divide the class. The young Alexander started to deliver a repeat lecture in the evening, and in 1754 his father requested that he be made 'Conjunct' (joint) professor, which was agreed. Professor Alexander Monro secundus, aged 21, had not yet graduated in medicine, which he did in October 1755, presenting a dissertation entitled *'De Testibus et Semine in variis Animalibus'*. He then also pursued further training in London (with William Hunter), Berlin, Paris, and Leiden. He returned to the University of Edinburgh in 1757 and from the next year, and for the following 22 years, delivered the full course of lectures in anatomy and surgery. During the long winter session he lectured daily from 1 to 3pm and spent every morning preparing anatomical specimens for his class. His lectures were praised for their clarity and cogency, and he became the most influential anatomy professor in the English-speaking world.

His particular interest was in the nervous system and the brain. He published on the communication mechanism of the lymphatic system, correctly defining its separateness from the circulatory system (and disputing the precedence of the discovery with, among others, William Hunter). The effect of opium and 'metalline' drugs on the nervous system was described in a later paper *'Experiments on the nervous system...'*. A more significant medical legacy comes from his investigation of the communication between the lateral ventricles of the brain, first published in 1783. In *'Observations on the Structure and Functions of the Nervous System'* he identified that the short connecting conduit now known as the 'foramen of Monro' is quite small in a healthy brain, but when an abnormal accumulation of cerebral fluid is present (hydrocephalus) the opening may be as large as one centimetre. Most significantly, he had the breakthrough realisation that a healthy brain cavity is rigid and of constant volume - and that this had important consequences. In his own words: *"For, as the substance of the brain... is nearly incompressible, the quantity of blood within the head must be the same, or very nearly the same, at all times, whether in health or disease, in life, or after death, those cases only excepted in which water or other matter is effused, or secreted from the blood vessels; for in these, a quantity of blood, equal in bulk to the effused matter, will be pressed out of the cranium."* The full implications of this were realised by his former student George Kellie.

Kellie was born near Edinburgh in the sea-port of Leith, where his father (also George) practised as a surgeon. Kellie junior followed his father's profession and after a five-year apprenticeship, became a surgeon in the Royal Navy, serving in France in the build-up to the Napoleonic Wars. Returning to Scotland in 1802, he became Surgeon to the Royal Leith Volunteers and a Fellow of the Royal College of Surgeons of Edinburgh. He qualified for his MD from the Edinburgh Medical School (with a dissertation on *'De Electricitate animale'*), was married, and established a successful surgical practice.

On the morning of Sunday, November 4th 1821, Kellie was asked to examine the bodies of three people, two men and a woman in her sixties, found lying outside near Leith after a violent, snowy overnight storm, and to determine the cause of the deaths. On autopsy of one of the men and the woman, Kellie found that the brains of the deceased were normal, except that the veins on the surface and meninges were congested with blood, while the arteries were bloodless. After eliminating Saturday-night drunkenness as a major factor, he concluded that the deaths resulted from cold and exposure, which can result in shrinkage or compression of the cranium, and 'disordered cerebral circulation'. He wrote: *"... in most cases of intrusion when the cavity of the cranium is encroached upon by depression of its walls, compensation may be made at the expense of circulatory fluid within the head; less blood is admitted and circulated...The argument has already been taken up and illustrated by Dr. Abercrombie...to shew the improbability of any intrusion being made on the brain without a corresponding displacement of some portion of its circulating blood."* Kellie went further and made this idea even more explicit: *"... if the total quantity of circulating fluid within the head be Z, and the quantities contained respectively in the arteries and veins be X and Y, then $X+Y=Z$. If now, the circulation become deranged...and if the surcharge 'a' become permanently congested in the arteries, the accumulation in those vessels will now be $X+a$, and the contents of the veins $X-a$."* He noted *"One of my oldest physiological recollections...is of this doctrine having been inculcated by my illustrious preceptor in anatomy, the second Monro...which he used to illustrate by a hollow glass ball, filled with water."* Kellie went on to confirm the hypothesis of constant intracranial fluid volume (now understood to include the volume of cerebro-spinal fluid, CSF) in a series of gruesome experiments in the bleeding of sheep and dogs. The insights and hypothesis continued to be promoted by the famous Edinburgh physician, John Abercrombie - the 'Dr Abercrombie' acknowledged by Kellie - particularly in Abercrombie's 1828 publication of *'Pathological and Practical Researches on Diseases of the Brain and Spinal Cord'* which is now regarded as the first textbook on neuropathology. The 'Monro-Kellie doctrine' became an accepted part of medical knowledge and underpinned the development of modern practices for the management of intracranial pressure.

George Kellie was elected a Fellow of the Royal Society of Edinburgh in 1823 and became President of the Edinburgh Medico-Chirurgical Society in 1827 - his successor to that office was John Abercrombie. Kellie collapsed and died in Leith in 1829 while on his way home from visiting a patient.

The second Alexander Monro continued his influential teachings and demonstrations at the Edinburgh Medical School until 10 years before his death, at the age of 84, from a brain haemorrhage. His son, Alexander Monro *tertius,* had become conjunct Professor of Anatomy and Surgery at Edinburgh in 1798. He was not to be as distinguished or popular as his illustrious forebears. He was an uninspiring lecturer and became embroiled in the body-snatching scandal of Burke and Hare in 1828. Perhaps the most famous Edinburgh student of Alexander *tertius* was Charles Darwin, who was disgusted by his dishevelled, dirty and sometimes bloody appearance and said of him: *"I dislike [Monro] and his lectures so much that I cannot speak with decency about them. He is so dirty in person and actions..."* and *"made his lectures on human anatomy as dull as he was himself."* Alexander *tertius* resigned his Chair in 1846, ending 126 years of continuous Monroe professorships at the University of Edinburgh.

18 The Colossus of roads: John Loudon McAdam (1756-1836)

Well-constructed roads were introduced into Europe by the Roman Empire and little changed in the following centuries. There was no single standard construction, but usually a Roman road was constructed on top of an earthen embankment and consisted of four layers placed on compacted soil: a foundation layer, often of stones of around 10cm diameter; a second layer of rubble, coarse sand and gravel, compacted with lime cement to form concrete; a 'nucleus' of fine cement of lime and pottery shards; and a surface layer of finer gravel or occasionally of paving stones.

Illustration 20: John Loudon McAdam - engraving by Charles Turner

Depending on the importance of the roads, their widths could be up to 12m (40ft), and their construction was of course very manually intensive. These methods were still unsurpassed in Britain until they were revamped by a self-taught Scottish engineer and businessman, John Loudon McAdam, who invented a new, economical road-building method.

McAdam was born into a family of some nobility in Ayr in 1756. His mother, Susannah Cochrane, was niece to the 7th Earl of Dundonald and his father James was Baron of Waterhead, and a minor laird. But in John Loudon's early years, the family was dogged by misfortune. Their home was destroyed by fire, their ancestral estate was sold, and his father died when McAdam was aged 14, leaving behind only a founding share in an Ayrshire bank which failed two years later. Forced to earn a living, the young John Loudon left home for New York in 1770, to join his aunt Ann and uncle William McAdam. William was thriving in trade and society, as a founder of the New York Chamber of Commerce and a principal officer of the St Andrew's Society of New York State. Young John Loudon showed a precocious aptitude for business, working lucratively as a 'prize-agent' in his uncle's premises; that is, procuring and delivering monetary rewards to ships' crews for capturing or sinking an enemy vessel - the money coming from the sale of the ship and cargo, or from the British Crown if the vessel was a captured warship. The outbreak of the American War of Independence in 1775 made this business untenable for the Royalist McAdams, and in fact William was to lose most of his fortune by confiscation after the American Revolution. John Loudon, however, managed to continue in business and was financially secure enough by 1778 to marry Gloriana Margaretta Nicoll, the daughter of a prominent and wealthy lawyer in the Colony. By 1783 the former American colonies had won their independence. John Loudon also had much of his fortune confiscated, and with Gloriana he sailed home from America with their two young children, to buy and settle in a house and small estate at Sauchrie, near Maybole in Ayrshire.

In returning to Ayrshire, McAdam became something of a country gentleman and entrepreneur. He was a magistrate and a deputy lieutenant for the county. He operated the local Kaims Colliery which supplied coal to the British Tar Company (owned by his relation Archibald Cochrane, the 9th Earl of Dundonald), and ran its kilns in Muirkirk. A slightly strange boyhood interest in roadways was to resurface too: he became a trustee of the Ayrshire turnpike roads, which were constructed mainly of loose gravel and were in a state of chronic disrepair. The British Tar Company was the first to produce coal tar commercially and profitably, and McAdam became its manager, and then acquired ownership of its patent rights and the Muirkirk works. The main product was coal tar for the coating of ship hulls, with a by-product, coke, for the nearby Muirkirk Iron Works. Sales problems arose however, and financial crisis struck again. The Sauchrie estate was sold and in 1798 McAdam, at the age of 42, moved his family south to settle in Bristol, where their seventh child, John Loudon junior, was born. The British Tar Company survived, and McAdam retained his financial interest in it, with his eldest son William installed as manager.

Bristol was a major port and rife with opportunity. McAdam, as always, threw himself into civic and business life. A spell working again as a prize agent in Falmouth was ended by the Peace of Amiens in 1802, which brought a temporary halt to Britain's war with Napoleon and post-revolutionary France. In Bristol, the family rented several comfortable houses in succession, and McAdam's second son James established a manufacturing company of 'Lampblack, Mineral Paints, Oils etc' exploiting the relationship with the British Tar Company. John Loudon's interest and enthusiasm for the much-needed improvement of roads was rekindled again and he became a trustee of the Bristol Turnpike Trust, which was responsible for around 150 miles of roadways.

McAdam wrote an extensive memorandum to Sir John Sinclair, chairman of the Board of Agriculture, that appeared in 1810 in the report of the parliamentary Select Committee on turnpike roads and highways, and again in 1816 in McAdam's first book, *Remarks on the Present System of Road Making*. He pointed out the mistaken policy of trying to regulate traffic to suit the poor condition of roads, rather than improving road construction to accommodate the increasing traffic loads. He specified a new method of road construction based on stones of carefully measured sizes and asserted that massive foundations of stone (as used by his compatriot and contemporary Thomas Telford) were unnecessary. The bottom layer, of thickness around 8 inches (20 cm), was to consist of stones up to 6 ounces in weight and 2 inches (5 cm) in diameter. The top layer (of thickness 2 inches) used stones of up to 0.8 in (2 cm) size. The size of stones was checked by supervisors carrying scales and workmen could check the size of the base layer stones by testing whether they could fit into their mouths. McAdam stressed the importance of the limit to the size and weight of the upper layer stones, so that they were much smaller than the 4-inch width of iron carriage tyres. No water-absorbent materials could be used in the construction, and ironically, no tar binding agent was to be used either. The surface, composed of angular, carefully-sized stones, was compacted into a level, solid, waterproof surface by the action of road traffic, or later by compaction by cast-iron rollers. A slight camber was specified to ensure that rainwater drained quickly off the road and did not penetrate the foundations. Even the tools to be used were carefully specified: small hammers to be about one pound in weight with faces the size of a shilling coin; rakes to have teeth two-and-a-half inches long. These simple but exact construction principles found wide acceptance and in 1815 the Bristol Turnpike trustees invited McAdam to turn his theory into practice, and to become Surveyor-General for the Turnpike Trust at a salary of £400 per year.

The road from Bristol towards the town of Shepton Mallet was successfully 'macadamised', and McAdam, with his sons James and William, went on to be surveyors or consultants to 25 Trusts by 1819. The same year McAdam published a pamphlet, called *'A Practical Essay on the Scientific Repair and Preservation of Public Roads'*, which explained his method, and was printed and distributed by order of the British government's Board of Agriculture and Internal Improvements. McAdam wrote in his introduction: *"The turnpike roads of England and Wales amount to about 25,000 miles in extent, and cost annually about a million and a quarter to maintain them, in their present defective state ...The following plans for the construction and repair of the roads of Great Britain, and the protection of the funds from mismanagement and speculation, were the result of the reflections of thirty years."* As well as the engineering specifications, McAdam (and the Government) were much concerned with the economics of road construction, and avoidance of the fraudulent accounting which was endemic in the Turnpike Trusts. McAdam wrote: *"The price of lifting a rough road, breaking the stones, forming the road, smoothing the surface, cleaning out the watercourses, and replacing the stone, leaving the road in a finished state, has been found in practice, to be from one penny to two pence per superficial yard, lifted four inches deep...At two pence per yard, a road of six yards wide will cost, therefore, one shilling per running yard or 88 [pounds] per mile."* In a revealing comment on the working practices of the time, McAdam also noted another advantage of breaking stones to his specified, smaller size: *"The Commissioners at Bristol used to pay fifteen pence per ton for limestone from Durdham Down, for the use of their roads, and broken to a size above twenty ounces: stone is now procured from the same place, broken so as none exceed six ounces, for ten pence per ton, and the workmen are very desirous of contracts at that rate, because the heavy work is done by the men, the light work with small hammers by the wives and children, so that whole families are employed."*

In 1823 McAdam petitioned parliament to apply his system of road-building throughout Great Britain. An enquiry by a parliamentary commission revealed the astonishing extent of his efforts to gather information and to perfect his methods. Between 1798 and 1814 he had travelled 30,000 miles and spent 2000 days examining the condition of British roads, at a personal cost of £5000, and had undertaken numerous construction experiments, also at his own cost. The committee reported favourably, eventually resulting in an award of £6,000 to indemnify his personal expenditure, and in 1827 he was made General Surveyor of Roads for England, Scotland and Wales. By then, 'Macadamed' roads were already spreading, first across Bristol and southwest England, then across Great Britain and abroad.

McAdam's international fame and commitments led to difficulties in Bristol. There the Turnpike trustees were feeling neglected. To placate them, McAdam appointed his son, John Loudon junior, as joint surveyor to the Bristol Trust to cover his own absences, which had provoked a vote of censure. McAdam senior had to resign, although he continued to consult for the Bristol Trust while surveying for the adjacent Bath Trust in partnership with his grandson William.

McAdam's wife Gloriana died in 1825, further weakening his ties to Bristol and the West Country of England. Two years later he moved to Hertfordshire and later married Gloriana's younger relative, Anne Charlotte Delaney. His surveyorships in road-building trusts continued to multiply, particularly in the north of England and in Scotland. He died on a visit to Scotland, in Moffat, aged 80, in 1836.

During McAdam's lifetime his road-building methods had become internationally renowned and widely adopted. The first macadamed road in North America was a 10-mile stretch completed in 1823 between Boonsboro and Hagerstown in Maryland. A 73-mile length of the National Pike (Cumberland Road), also in Maryland, was rebuilt and macadamed, completing in 1830. By the end of the 19th century, most of the main roads in Europe had been macadamised. Further construction improvements of course ensued. With the development of fast-moving motor vehicles, dust clouds and scattering of gravel surfaces became a problem. The solution of spraying a layer of coal tar to bind the surface (originally resisted by McAdam) became known as tar-bound macadam and in 1902 the Welsh inventor, Edgar Purnell Hooley, patented a mixture of tar and slag aggregate that he called tarmacadam or 'tarmac'.

McAdam's sons and grandsons continued in their father's profession. His second son, James Nicholl McAdam, took the surveyorship of the Epsom Trust in 1817 at the request of his father and went on to work for 39 turnpike trusts including the huge London Metropolitan Trust. He displayed the same professional talents in road construction and financial management as his father, and was knighted in 1834 (his father having declined the same honour).

A famous 1827 cartoon by Heath depicts a giant kilted McAdam straddling the great road to London clutching money bags under each arm and is titled *'Mock-Adam-izing - the Colossus of Roads'* (a pun, of course, on the Colossus of Rhodes). The same jokey title was sometimes bestowed on James and on the McAdams' contemporary civil engineering rival, Thomas Telford. But it is for John Loudon McAdam's singular, intense focus on the engineering, process and cost-control of road-building that the title is truly deserved.

19 Master of civil engineering: Thomas Telford (1757 – 1834)

Illustration 21: Thomas Telford - engraving from a portrait by S. Lane

Civil engineering as we know it today was in its infancy in the year of Thomas Telford's birth. Most constructions were designed in timber, brick or stone, and most designs were the result of long experience, skilled craftsmanship, and trial-and-error. In Telford's lifetime, much was to change, and he was the driver of many of the changes that made possible the modern discipline of civil engineering.

Telford was born in the rural parish of Westerkirk in Eskdale, Dumfriesshire, the second son of Janet Jackson and her husband, a shepherd, John Telford, who died 4 months after Thomas was born. Like so many of the luminaries of the Scottish Enlightenment period, Telford was raised by his mother, and given a good basic school education. The family was poor, and needed the support of Janet's brother, after whom Thomas was named. Telford remained at the local parish school until the age of 14, when he was apprenticed to nearby stonemasons, constructing houses and small bridges for the local farmers. The constructions were simple, but the experience was wide-ranging. As Telford was to write in his autobiography: *"...peculiar advantages are thus afforded to the young practitioner; for as there is not sufficient employment to produce a division in labour in building, he is under the necessity of making himself acquainted with every detail in procuring, preparing and employing every kind of material."* Langholm Bridge, completed around 1778, is said to be the first to bear Telford's stonemason's mark.

Eskdale could never provide the young Telford with challenges suited to his talents and ambition. Aged around 23, he moved to Edinburgh and observed the construction of its Georgian 'New Town' while absorbing the lessons of its architecture and learning draughtsmanship. Two years later he moved on to London and found employment as a stonemason working on the reconstruction of Somerset House, under chief architect Sir William Chambers, to provide for its occupation by the Royal Society and the Royal Academy of Arts. Telford wrote *"...I became known to Sir W. Chambers and Mr. Robert Adam, the two most distinguished architects of that day; the former haughty and reserved, the latter affable and communicative...although from neither did I derive any direct advantage ...the latter made the most favourable impression, while the interviews with both convinced me that my safest plan was to endeavour to advance, if by slower degrees, yet by independent conduct."*

At this point, surprisingly, Telford's upbringing in rural Eskdale provided him with an influential new patron. He was asked by the soldier and liberal politician, Sir James Johnstone, to advise on work on his Eskdale property, and was instructed in the details by Johnstone's brother, William Pulteney, husband of the heiress of the earl of Bath. Pulteney (born in Eskdale and called William Johnstone until he married the heiress Frances Pulteney and changed his name) was an advocate and Whig member of parliament for Cromarty and then Shrewsbury. He was reputedly the richest man in Britain; and in his Edinburgh days as a younger man, he had been acquainted with David Hume, Adam Smith, and indeed, the architect Robert Adam. Pulteney liked Telford's work and approach, and employed him to restore Sudborough Rectory in Northamptonshire. This was the beginning of a close friendship and many more commissions from the wealthy MP for Shrewsbury. Telford was invited to Shropshire to restore Shrewsbury Castle as an occasional home for Pulteney, then supervised the construction of the new county gaol, and became county surveyor of all public works. Such was Telford's close involvement in these projects that in Shropshire he became known as 'Young Pulteney'. His patron William, the 'old' Pulteney, was in 1790 appointed Governor of the British Fisheries Society and commissioned Telford to design many fishing port harbours and dwellings, including those at Tobermory, Ullapool and the largest, at Wick, where 'Pulteneytown' was planned and developed over several decades. (The excellent and renowned 'Old Pulteney' single-malt whisky is still distilled in Wick). At the aborted harbour development in Lochbay, Skye, Telford used the recently-patented 'Roman cement', a fast-setting artificial mixture of clay and chalk which was highly water-resistant and used subsequently by Robert Stevenson in lighthouse construction.

In 1793 Telford was appointed 'General Agent, Surveyor, Engineer, Architect and Overlooker' to the construction of the Ellesmere Canal, connecting the rivers Severn, Dee and Mersey. In Shropshire, Telford had worked on the Shrewsbury Canal and designed the world's first long cast-iron aqueduct (57m, 186ft) across the River Tern at Longdon. He used this experience at Pontcysyllte to carry the Ellesmere Canal across the River Dee on a 307m length of cast-iron aqueduct supported on 18 stone and iron pillars. This magnificent aqueduct took 10 years to build, cost £47,000 (equivalent to around £4 million in 2020) and it remains the longest aqueduct in Great Britain and the highest in the world. Telford's work on canals continued, initially with his Ellesmere Canal co-designer William Jessop, on the design of the Caledonian Canal through the Great Glen of the Scottish Highlands. This mammoth project required many engineering innovations including the use of steam engines for pumping and dredging, and the construction of 28 locks, 30m wide. After much delay, a reduction in the planned depth of the canal, and a huge cost overrun to £982,000, twice that budgeted, the canal was finally completed and opened in 1822. The Napoleonic Wars were partly responsible for the problems. Telford wrote later about the difficulty of the project. *"The chief cause was the unprecedented warfare in which all Europe was involved during the times the work was in progress; the value of materials and labour rising from 30 to 50 percent, so that the sum annually granted remaining the same, only one-half the quantity of work could be annually performed.*" Telford went on to be one of the great canal builders of the Industrial Revolution - working on at least 33 literally ground-breaking canal projects in Britain as well as consulting on the Gotha Canal in Sweden.

Building canals and aqueducts was just a part of Telford's wide engineering career. Over his lifetime he designed, built or consulted on thousands of stone bridges including 1100 bridges to a standard specification in the Scottish highlands, and the elegant bridges at Dunkeld, at the Broomielaw in Glasgow, and the Dean Bridge in Edinburgh. His pioneering use of cast-iron was employed in the Buildwas Bridge over the Severn and his elegant prefabricated spandrel arches can be seen still at Holt Fleet Bridge on the Severn, Craigellachie Bridge on the Spey, and Galton Bridge in the West Midlands. His biographer Tom Rolt has said of Telford: "*No other man has ever handled cast iron with such complete assurance and understanding, his exact knowledge ... enabling him to achieve that perfection of proportion which gives strength the deceptive semblance of fragility.*"

The Menai Strait between Bangor and the island of Anglesey in northwest Wales presented Telford with a major challenge which resulted in arguably his greatest achievement. The high banks, strong winds and fast-flowing currents of the 16-mile-long Strait made for a difficult crossing for ferries, and the route was increasingly important to serve Anglesey's port of Holyhead and its shipments across the Irish Sea to Dublin. In 1819 an Act of Parliament was passed commissioning a bridge across the Strait, the construction of a new road across Anglesey to Holyhead, and improvements to the road along the North Wales coast from Bangor to Chester. Telford later noted that: "*On the Irish side...between the landing place at Howth and the city of Dublin, was a very imperfect road, which, under the authority of the Commissioners, has been rendered in all respects equal to the road in North Wales*". Ireland was at that time, of course, part of the United Kingdom since the Acts of Union of 1800.

For the Menai Bridge, Telford proposed a radical bridge design using new technology: a bridge suspended from wrought-iron chain cables, with an unprecedented span of 176 metres. Telford had experimented with designs for a iron-cable suspension bridge at Runcorn in 1814 and his Menai proposal was accepted, but it was bold. Telford was aware of the cable-stayed footbridge at Dryburgh Abbey (between Dryburgh and St. Boswells in the Scottish borders) which collapsed in 1818. More encouragingly, the Union Bridge across the River Tweed between Fishwick in Scotland and Horncliffe in England was based on a chain suspension design of which Telford knew and approved. It was designed by a Royal Navy officer, Captain Samuel Brown, son of a Galloway man, William Brown of Borland. Captain Brown had introduced the use of wrought-iron chain cables to the Navy, and took out patents on wrought-iron chain-making in 1816 and 1817, going on to found a company that supplied all the chain to the Navy until 1916. The Union Bridge construction began after work started on the Menai Bridge, and its span of 137m was completed in less than 12 months in July 1820. The project was mentored by John Rennie (the elder), the eminent Edinburgh civil engineer who had worked as a young man for Boulton and Watt, and who had proposed two designs for a cast-iron bridge over the Menai Strait in 1801. On its completion the Union Bridge was the world's longest suspension bridge, and it is the oldest suspension bridge still carrying road traffic.

The challenges of the Menai Strait meant that construction of Telford's suspension bridge took several years. Work began in 1819 and the limestone piers were completed in 1824. The massive wrought-iron suspension chains were produced by Hazeldine's factory in Shrewsbury and rust-proofed with warm linseed oil and paint. The chains and ironwork weighed nearly 2,200 tons. A coat of the paint weighed 2 ½ tons. The 16 chains were hauled, one-by-one, to the tops of the towers by 150 men using block and tackle, and anchored into solid rock on either shore. Finally, the bridge was opened in January 1826. The first coach to cross was the Royal London to Holyhead mail coach, which had planned to use the Menai ferry as usual, but was diverted and crossed the windswept bridge at 1.35 a.m. on the stormy night of 30th January. It would never use the ferry again. The next day the bridge was crossed by a succession of private coaches of the gentry and political supporters of the project, and it became the venue for a celebratory party. At the age of 68, Telford's genius and reputation as the finest civil engineer of his age was confirmed. His own account of the achievement, 8 years later, was typically modest and understated: *"Thus was successfully accomplished a complicated and useful bridge of unexampled dimensions, which has...converted what was formerly a disagreeable and sometimes dangerous journey ...into an object of national curiosity and delight."*

Telford's legacy extends far beyond his constructions of major canals and innovative bridges. His aptitude, versatility and self-taught skills in civil engineering were applied to a huge range of public projects. In addition to designing the remote harbours for the British Fisheries Society mentioned already, he engineered improvements to more than 100 docks and piers including those at Aberdeen, Glasgow, Leith, Dover, and St. Katherine's Dock in London. For the Highland Churches Commission set up by Act of Parliament, he oversaw the construction of churches and manses from Islay to Shetland, to standardised designs, which were known as 'Telford Churches' or 'Parliamentary Kirks'. With his fellow Scottish engineers, the John Rennies (both elder and younger) he advised on the construction of the 'cuts' that drained the English fens in East Anglia. From 1806 until 1822 he was engineer for the Glasgow waterworks, working with James Watt, and consultant on the construction of Glencorse Reservoir and dam to supply water to Edinburgh. He engineered major new arterial roads from London to Holyhead, from Glasgow to Carlisle, and advised Tsar Alexander I, on the improvement of the road from Warsaw towards Moscow, after Russia's successful defeat of Napoleon's army and its annexation of eastern Poland. His long-standing friend, the poet laureate Robert Southey, dubbed Telford 'the Colossus of Roads', a sobriquet that was also applied to John Loudon McAdam and his son James. Perhaps the pun was too good not to be used again and again.

Telford was the first President of the Institution of Civil Engineers from 1820 until his death, and was elected to Fellowships of the Royal Society and the Royal Society of Edinburgh. He died in 1834, unmarried, with no known surviving close family but with many close friends who remembered his warmly benevolent character. He is buried in Westminster Abbey and honoured in the name of Telford new town in Shropshire and its associated high school and technical college. He was a well-read man of letters too, and wrote many poems under the pen-name 'Eskdale Tam', including an elegy on the death of Robert Burns. One verse in that poem could equally apply to the self-taught young Telford himself:

"Nor pass the tentie [shy] curious lad,
Who o'er the ingle hangs his head,
And begs of neighbours books to read;
For hence arise,
Thy country's sons, who far are spread,
Baith bold and wise."

20 Rainwear from raintown: Charles Macintosh (1766-1843)

Glasgow is a rainy place. Not only was Glasgow the rainiest seat of learning in 18th century Scotland, it consistently ranks as the rainiest city in Great Britain, uncomfortably exceeding one metre of rainfall every year. Perhaps this was the motivation for Charles Macintosh to think about how to stay dry, and to use his spare time to do something about it. Charles was born in Glasgow in the (no doubt rainy) mid-winter of 1766, to a merchant family, and educated there at the grammar school.

Illustration 22: Charles Macintosh portrait by John Graham Gilbert

His proclivity for Latin in particular marked him out as a bright scholar and he was sent for a few years to Catterick Bridge school in Yorkshire, where he seemed to be as much teacher as pupil. After school he was apprenticed as an accountant to the company of John Glasford of Dougalston, a prominent Glasgow merchant. Glasford (nowadays spelled Glassford) was a Tobacco Lord - the name given to the prosperous Glasgow merchants who made huge fortunes importing the leaves from British colonies in the West Indies and Virginia. The trans-Atlantic westerly Trade Winds which deliver Glasgow's rainclouds also gave its traders a two- or three-week advantage over rival ports in Britain and Europe. In 1747 the French monarch, Louis XV, had granted Glasgow a monopoly for the exportation of tobacco to French ports. The result was a thriving triangular trade operated shamelessly on the backs of slave labour. Tobacco, sugar, and cotton were imported from the Americas to Europe; manufactured goods, textiles and rum were ferried to West Africa; and slaves were transported to the West Indian and American colonies. Glasgow's economy boomed, and the Tobacco Lords were the new aristocrats. John Glasford was one of the wealthiest of all, and was said at one time to have owned 25 ships and their cargoes, and traded around a half million pounds sterling in a year.

Charles Macintosh's interests were very different, however. For him, the future lay in science, and in particular, chemistry. He attended the lectures of Joseph Black in Edinburgh, and of Black's protégé William Irvine, in Glasgow. In 1786 he started in business with his father, producing sal ammoniac, (ammonium chloride) principally for use in the tinning of iron, copper and brass, and began manufacturing Prussian Blue dye and sugar of lead (lead II acetate, used in textile printing and dyeing) as well as acetate of alumina, which has medical applications. While still only 20 or 21 years old, he was also active in the Glasgow Chemical Society and in the Glasgow Commercial Society, contributing opinionated essays on scientific and political subjects including the promotion of woollen manufacture in Scotland, the prospects for uniting Ireland and Great Britain, and commercial relations between Britain and other European states. In 1790 he married Mary Fisher, the daughter of another Glasgow merchant, Alexander Fisher, and they had a son, George, named after Charles' father.

Around this time Macintosh found a use for the shale waste from nearby exhausted coal mines, as a source of potash alum (potassium aluminium sulphate) used for fixing dyes in textiles, and in the tanning of leather. Seven years later Macintosh created Scotland's first alum works near Paisley. He subsequently built a highly successful and lengthy business career, inventing many new commercial processes for the production of useful chemicals, including a lucrative method of making the bleaching agent chloride of lime (though the patent was filed in the name of Charles Tennant, a business partner in the St Rollox chemical works). It was in 1820, while experimenting with coal-oil (naphtha), a by-product of coal gas production, that Macintosh made the invention which made him famous. Using coal-oil as a solvent for natural rubber, he cemented two layers of cloth together with a layer of rubber, to create the world's first waterproof fabric capable of commercial manufacture. The patent for the process was obtained in 1823. The same year, Macintosh was elected a fellow of the Royal Society and he established a manufacturing and sales business in Glasgow, and in its rival for the claim to be Britain's rainiest city, Manchester. A wide range of waterproof items was soon available and exported worldwide, especially raincoats, the archetypal Macintosh - nowadays spelled Mackintosh - product. Macintosh soon diversified into other rubber-based products: tyres, air-beds, pillows and carry-bags for Arctic expeditions. Initially there were problems with the stiffness and durability of the rubber, especially in very cold or hot conditions. These were overcome by one of Macintosh's business partners, a self-taught manufacturing engineer from Wiltshire called Thomas Hancock, who shared Macintosh's interest in waterproof fabrics, initially aiming to protect passengers travelling in the coaches Hancock built with his brother, in his first business venture. In 1825 Hancock patented a

Illustration 23: An advert for Macintosh rainwear from Carson, Pirie & Co's 1893 catalogue

process for making 'artificial leather' using shredded rubber dissolved in coal oil and turpentine, probably inspired by Macintosh's patent of two years before. Macintosh and Hancock had merged their businesses to form Macintosh & Co., and Hancock became a full partner in 1834. Hancock invented and introduced a process to 'vulcanise' the rubber by heating it in the presence of sulphur, which cross-links the rubber molecules and improves the strength and flexibility of the material. Vulcanisation of rubber was developed independently by Hancock, who filed a patent in November 1843, and by the American, Charles Goodyear, who filed a U.S. patent on the process 8 weeks later in January 1844. The priority of the invention was of course disputed. In any case, vulcanisation solved Macintosh's problem with fabric stiffness and Hancock went on to be the principal driver of the British rubber industry.

Macintosh's patent for waterproof fabric was infringed in 1836 by a London firm of silk merchants called Everington & Son. Macintosh litigated and the resulting court case created even more publicity for his products. Before the Lord Chief Justice Tindal, the Attorney-General opened the proceedings by addressing the jury: *"You, gentlemen, have to decide whether Mr. Macintosh's patent for making Macintosh's waterproof cloaks be valid or invalid. Gentlemen, you have often heard of these cloaks... [which] have obtained the greatest celebrity. Gentlemen, this patent was taken out in the year 1823; it was respected until within the last few months: it has lately been infringed by the defendants, Messrs. Ellis and Everington, and one or two others, who are acting in concert with them. I think that it is a strong presumption in favour of its validity, that no one has called it in question for a considerable series of years, and not until it is within two years of the period of its expiration."* By now the jury must have been gleaning the attorney's thinking on the matter. He went on: *"Mr Macintosh...applied a cement made of Indian rubber, which is brought from South America, the native name of which is caoutchouc- I may just as well call it Indian rubber, if you please...It could then be applied to various purposes, such as making flexible air-tight and water-tight vessels and tubes; and above all to cloaks, which are light, flexible, comfortable and entirely waterproof."* Jury direction and advertising endorsements complete, the attorney produced a series of chemists and employees of Macintosh & Co. to testify that the process used in Macintosh's factory was the same as the process described in the patent. In fact, inevitably improvement and changes had been made over the course of time, and unpatented process improvements had been made, by Hancock in particular. Anyway the jury found in favour of the patent's validity, and the 'Caoutchouc Patent Case' only added to Macintosh's fame. He chose not to extend his patent, since the many manufacturing improvements had superseded the original process.

In his later years Macintosh continued to promote the rainwear business, travelling extensively in Europe, corresponding in English and French, and maintaining his interests in his other early inventions, notably his 1825 patent for converting iron to steel by heating in the presence of carbon from coal gas. He went into business partnership with his fellow Glaswegian, James Beaumont Neilson, who developed the 'hot-blast' process for improving the efficiency, quality and yield of iron manufacture. Neilson's invention was simple, consisting in merely heating the air blown into the smelting furnace in a separate pre-heater, but this overthrew the accepted wisdom of the time which was that the blast should be as cold as possible. Heating the blast air enabled cheaper fuels to be used to fire the furnace and improved fuel efficiency. Macintosh helped Neilson to patent the process and the amount of iron produced using Neilson's process grew rapidly. By 1840 almost all the furnaces in Scotland were using hot-blast and in England and Wales, 58 ironmasters had taken licences at one shilling per ton of production, giving an annual income to Neilsen and partners of £30,000 per year. Neilson's lucrative patent (in which Macintosh had been awarded a share) was due to expire in October 1842, with the possibility of an extension, and its validity was challenged. Again Macintosh was involved in patent litigation, to defend his partners' interests. At one point the patentees pursued 20 simultaneous court actions against various companies in England and Scotland. The first case was brought to court in 1841 against the Harford Company in England. The verdict found in favour of the patentees. The next year the case against a group of Scottish ironmasters came to trial and again found in favour of Neilson and his partners, but went to appeal in the House of Lords on a technicality. During the delay, the patent expired and the 1843 case of 'Neilson vs Baird' in the Edinburgh court of session became improbably famous as 'the great hot-blast affair'. The trial lasted 10 days and called 102 witnesses, in an action costing over £40,000. Summing up, Lord Justice General Boyle found the patent valid and awarded the patentees £7000 in damages and £4,867-16s in lieu of profits. Neilson and his partners continued to pursue Baird and claimed £20,000 in damages and £160,000 in lieu of profits lost from May 1840 to the patent's expiry. Baird ended up settling out of court in 1844 for a total of £160,000 in damages and lost profits.

Macintosh did not live to see the final act in the 'great hot-blast affair'. He died in July, 1843, a wealthy man survived by two of his three children, including George, who would be his biographer. His invention of the 'Mackintosh' was so successful that his name passed into the English language as the generic term for that not-so-simple invention - the waterproof raincoat. Manufacture of premium-priced rainwear for the international market still continues in the 'Mackintosh Ltd' factories at Nelson, Lancashire, and Cumbernauld in Scotland.

21 Enlightenment lights: Robert Stevenson (1772-1850)

The first mention of a lighthouse in Scotland was in 1635, when a patent was granted by Charles I to erect a light on the Isle of May at the mouth of the Firth of Forth. The duty of a penny ha'penny per ton for Scottish vessels, and twice that rate for foreign ships, was to be levied for its maintenance. The Scots Parliament ratified the patent in 1641 and the method of lighting was by coal fire. The Board of Northern Lighthouses was established in 1786 and would eventually acquire the Isle of May and its lighthouse in 1814, for the sum of

Illustration 24: Bust of Robert Stevenson, lighthouse engineer, by Samuel Joseph

£60,000. In the intervening period the technology and engineering of lighthouse construction was revolutionised by one man and his dynasty of direct descendants.

Robert Stevenson was born in Glasgow in 1772, the only child of young parents plagued by misfortune in Robert's infancy. His father Alan, a partner in the West India trading house in Glasgow, died of tropical fever in St. Kitts while visiting his elder brother Hugh, who managed the business overseas. A similar fever killed Hugh, in Tobago, effectively ruining the business. Robert was only two years old. His father was only 22 years old when he died. His mother, Jean Lillie, the same age, was thrown into a desperate widowhood; both her parents had died the same year. She moved to Edinburgh to live with her married sister, and in due course Robert was sent to the charity school run by Edinburgh's Orphan Hospital. Jean remarried a Glasgow merchant, James Hogg, in 1777, and they had two sons. Hogg soon deserted her and left for England, leaving her again in emotional and financial distress, caring for young Robert, though not for his half-brothers, of whom there are no records. At age 14, Robert was apprenticed to an Edinburgh gunsmith; and during his apprenticeship Robert met the man who would have the greatest influence on his future career - Thomas Smith.

Smith was a shipowner, and had his own Edinburgh business as an iron merchant, tinsmith and lamp-maker. He had become interested in the opportunity to provide reliable illumination for lighthouses, and had developed an oil-fired reflector lamp to replace the coal fires or candles which were the usual sources. In early 1787 Smith was appointed as engineer to the Northern Lighthouse Board, which had just been formed to provide lights for the coast of Scotland and the Isle of Man. He was 35 years old, had been twice married and was twice a widower, with three surviving young children, Jane, Thomas and Janet, to care for. Robert's mother Jean had been friendly with both Elizabeth Couper and Mary Jack, Thomas Smith's deceased wives. By 1791 she had moved, with Robert, to Smith's household to care for his children, and Robert, aged 19, began working for Smith and the Lighthouse Board. Jean secured a divorce from James Hogg in 1792, and married Thomas in the same year. They would go on to have a daughter, Elizabeth, known as Betsy.

Robert's engineering education was incomplete. He enrolled in the very practical natural philosophy classes of 'Jolly Jack' John Anderson in Glasgow, and over the next two winters attended lectures there at what would shortly become the Andersonian Institution. Stevenson wrote: *"It was the practice of Professor Anderson kindly to befriend and forward the views of his pupils; and his attention to me...was of a very marked kind, for he directed my attention to various pursuits, with the view to my coming forward as an engineer."* In the summers, Stevenson worked for Smith on lighthouse lamp installations and new lighthouse constructions on Little Cumbrae in the Firth of Clyde and on the Pentland Skerries (near Orkney), which Stevenson supervised, requiring him to live for four months in a tent on an uninhabited island. On returning from the Skerries he had a narrow escape from death. Sailing home on the sloop *Elizabeth*, the ship was becalmed, and the captain set Stevenson ashore to continue his journey to Edinburgh by land. A gale drove the ship back to Orkney where she foundered with the loss of all hands.

Stevenson's next major project was the famous Cloch Lighthouse near Gourock on the Firth of Clyde. The white tower of the Cloch became a familiar landmark for generations of Scots as they sailed away as emigrants to foreign lands, or as its welcoming light greeted them as they returned home. Stevenson and Smith installed and lit its oil lantern in August 1797. By then Robert was working full-time for Thomas on Lighthouse Board projects, and their relationship became even closer in 1799 when Stevenson married his own step-sister Jane Smith, so that his step-father also became his father-in-law. By 1802 Robert was a full partner in Smith's business, and had celebrated the birth of his and Jane's first child, a daughter, Jean. Thomas Smith was nearing retirement, but he and Robert collaborated on two more lighthouse projects, at Inchkeith on the Forth near Edinburgh and, in 1806, at Start Point on the Orkney Island of Sanday, where the previous, inadequate unlit tower was replaced and for the first time (at least in Scotland) a revolving light was installed. The construction was difficult and visited by another tragedy. Workers who were building the lighthouse boarded the vessel *'Stromness'*, to sail home to Edinburgh. A violent storm blew up and the *Stromness* sheltered off the island of Flotta. During the night her anchor cables broke, and she was smashed onto rocks. All on board were drowned except the cabin boy, who was found clinging to the boat's mast.

Smith retired in 1808 and Stevenson was appointed as his successor as engineer to the Northern Lighthouse Board. He was 36 years old, and was also the father of two recently-born sons, Alan and Robert, who survived the prevalent threats of infant diseases and grew to adulthood. Stevenson's first completed project for the Lighthouse Board as its principal engineer was to become perhaps his most famous achievement. The Bell Rock reef stood in the North Sea some 12 miles south-east of Arbroath and presented a treacherous hazard to vessels sailing between the Firths of Forth and Tay. Also known as Inchcape (or Cape) Rock, Stevenson wrote of it: *"There is a tradition that an Abbot of Aberbrothock [Arbroath] directed a bell to be erected on the Rock, so connected with a floating apparatus, that the winds and sea acted upon it, and tolled the bell, thus giving warning to the mariner of his approaching danger... the bell, it is said, was afterwards carried off by pirates, and the humane intentions of the Abbot thus frustrated."* True or not, the Rock was a major danger to ships sailing along the east coast of Scotland and the need for a lighthouse was clear. A violent storm in 1799 wrecked around 70 vessels around the east coast and had prompted Stevenson to propose a beacon for the Bell Rock, supported on 6 cast-iron pillars. But Bell Rock presented a severe environment and serious structural challenges. Its surface is uncovered only at low tide and at high water is submerged to a depth of around 16 feet (5m). Stevenson's first visits to the sea-battered Rock, in October of 1800, convinced him that a cast-iron beacon would be too vulnerable, and that a stone lighthouse was needed, on the lines of John Smeaton's Eddystone Lighthouse near Plymouth. He built a scale model of his intended design and submitted his plan to the Northern Lighthouse Board, with a cost projection of 42,865 pounds and 8 shillings. The Board was sceptical, but in 1802 one of its Commissioners, Lord Hope, Member of Parliament for Edinburgh, raised a Bill in the House of Commons to authorise the work and enable the Lighthouse Board to raise coastal duties to pay for it. The Bill failed to pass through the House of Lords, with opposition from the Corporation of the City of London, worried that too long a stretch of coast would be included in the tariff charges. However, the need for action soon became even clearer. In 1804, *HMS York*, a 64-gun warship, foundered on the Bell

Rock, with the shocking loss of all on board. Pressured to act, the Northern Lighthouse Board asked for a second opinion from John Rennie the Elder, the eminent Scottish civil engineer and Fellow of the Royal Society, who was resident, and well-respected, in London. Rennie visited the Rock in August 1805 and concluded that the project was necessary and feasible. The next summer, the 'Cape Rock Lighthouse (Scotland) Act' was passed by Parliament. The Board paid Rennie the sum of £400 and at the end of 1806 gave him the title "Chief engineer for conducting the work". Rennie immediately asked for Stevenson (who had not yet succeeded Smith as the Board's principal engineer) to work as his assistant and to execute the work.

Rennie suggested to Stevenson some design modifications, also cognisant of the innovations of the Eddystone Lighthouse. Some were accepted by Stevenson and some were not. In any case they agreed to a taller and slimmer tower above a tapered base of cycloidal, dovetailed stone curves, to dissipate wave energy, and faced with waterproof 'Roman cement'. Rennie's projection of the cost was similar to Stevenson's at £41,843-15s, and their working relationship was a good one. Stevenson was under no illusions about the challenge. He wrote: *"All knew of the difficulties of the erection of the Eddystone Lighthouse, and the casualties to which that edifice had been liable; and in comparing the two situations, it was generally remarked that the Eddystone was barely covered by the tide at high water, while the Bell Rock was barely uncovered at low water."* Rennie wrote to encourage his young colleague, referring to Smeaton, who had died in 1792: *"Poor old fellow, I hope he will now and then take a peep of us, and inspire you with fortitude and courage to brave all difficulties and all dangers, to accomplish a work which will, if successful, immortalise you in the annals of fame."* Work began in 1807, the same year that Stevenson's eldest son Alan was born.

The work progressed using a leased yard in Arbroath for stone cutting and test fitting. An 81- tonne ship (captured from Prussia and renamed *Pharos*), was illuminated and moored near the Rock to provide material stores and worker accommodation; and a new 41-tonne sloop was built in Leith and named *Smeaton* by Stevenson, to provide ferry

Illustration 25: Bell Rock Lighthouse during a storm from the North-East - J.M.W. Turner

transport from shore. Work was constrained to around two hours at low tide, and all work had to be halted during the winter seasons. The foundation stones were sunk into the bedrock in mid-summer of 1808, at the level of the low water spring tide, and by the end of the season four courses of the tower were complete. Stevenson and his foreman, Francis Watt, devised balance cranes, moveable jibs and an elevated cast-iron railway to transport the materials around the Rock. John Rennie made an annual visit to advise and inspect progress, and Stevenson laid the last stone atop the tower in September 1810. By the end of that year the interior and light room were finished and the Bell Rock Lighthouse showed its light for the first time in February 1811. The illumination was by an array of seven 'Argand' high-brightness, low-maintenance oil lamps, in front of 24 parabolic silvered reflectors, with coloured lenses to display both white and red flashes of light. The whole arrangement was rotated by clockwork driven by a weight descending through the tower and maintained by three lighthouse keepers. The total design and construction cost of the project came to £61,331. The structure remains - and is the oldest sea-washed lighthouse in the world.

The success of the Bell Rock Lighthouse was the springboard for Stevenson's career in civil engineering. He designed 16 more lighthouses for the Northern Lighthouse Board as well as directing the installation of lights in Ireland, the Isle of Man and British overseas colonies; and as lighthouses proliferated, it became essential to enable navigators to distinguish between them. Stevenson developed further the idea of rotating lights and introduced lighthouses which flashed intermittently at recognisable intervals, recognised by the award of a gold medal from the King of the Netherlands. Robert established a highly successful civil engineering practice in Edinburgh and designed numerous harbours, canals (working with Telford), navigation channels, roads, bridges and railways. He was joined in building lighthouses for the Northern Board by his sons Alan and David and, after his retirement, they were joined by his youngest son Thomas, father of Robert Louis Stevenson, and the inventor of the 'Stevenson Screen' used in meteorology. His sons and their descendants would maintain the engineering practice continuously until 1952.

Robert Stevenson was chief engineer for the Northern Lighthouse Board until 1842, when he was succeeded by Alan. He died at home in Baxter's Place, Edinburgh in 1850. Stevenson's sons were fiercely proud of their father's achievements, and Alan argued jealously with John Rennie the Younger about the relative contributions of their fathers to the Bell Rock Lighthouse. Similarly Thomas disputed the precedence of the flashing light inventions of John Richardson Wigham in 1870. The arguments were unnecessary because Stevenson's reputation and legacy were secure. The last words can go to the Northern Lighthouse Board to which he devoted so much of his career. *"The Board...desire to record their regret at the death of this zealous, faithful and able officer, to whom is due the honour of conceiving and executing the great work of the Bell Rock Lighthouse...and will long be remembered by the Board."*

22 Through a lens brightly: Robert Brown (1773-1858)

Illustration 26: Robert Brown, botanist – from portrait by H.W. Pickersgill

'Situation Vacant - Naturalist wanted to join an expedition to New Holland' was an advertisement that never appeared in 1798, though it might well have done. The Scottish explorer of West Africa, Mungo Park, had withdrawn from a proposed expedition to the interior of New Holland, which we now call Australia, leaving a vacancy. Park's compatriot and former fellow student, the botanist Robert Brown, was immediately interested. He used his network of contacts to write to the expedition's lead naturalist, Sir Joseph Banks, a veteran of Captain Cook's first great voyage, seeking to be recommended. His letter of recommendation said *"Science is the gainer in this change of man; Mr. Brown being a professed naturalist. He is a Scotchman, fit to pursue an object with constance and cold mind."* Brown was not selected. The expedition was in any case curtailed, but a year later, the explorer Captain Matthew Flinders proposed a new expedition to Banks, one that would determine if New Holland was a single island or many. The proposal was accepted and in December 1880 Flinders offered the position of 'expedition naturalist' to Brown.

Robert Brown was born in Montrose, the son of Helen Taylor and James Brown, a minister in the Scottish Episcopal Church, whose Jacobite sympathies were so strong that he dissented to accept his church's allegiance to King George III. Robert attended Montrose Academy and Marischal College in the University of Aberdeen, before his father's dissention required the family to move to Edinburgh in 1790. Robert enrolled and studied medicine at the University of Edinburgh, but became fascinated by natural history and in particular, botany, making expeditions into the Highlands to assemble a herbarium of Scottish plants. He gave up his university studies in 1793 and enlisted in the 'Fife Fencibles', a kind of home defence unit that was nevertheless part of the regular army, which would have displeased his father – who had died two years previously. Posted to Ireland as a surgeon's mate, in the relatively peaceful years after the Roman Catholic Relief Act had given Irish Catholics the rights to vote and hold public office, Brown found plenty of time to continue his botanical studies. On leave, he met the Portuguese botanist José Francisco Correia de Serra, who was working for Sir Joseph Banks, and who was the referee who wrote to Banks recommending Brown for the abortive New Holland expedition.

Back in Ireland in 1799, Brown continued with his botanical fieldwork. He trained himself in the use of microscopes and focused on cryptogams, the families of plants that reproduce without seeds or flowers, including ferns, seaweeds, lichens and mosses, discovering several new species. (They are not to be confused with *cryptograms*, though amusingly the British government would make this mistake during World War II, when they recruited Geoffrey Tandy, a marine biologist and expert in cryptogams, to the cipher code-breaking station at Bletchley Park. The posting actually bore fruit when some unreadable, water-logged enemy codebooks were captured and sent to Bletchley. Tandy's experience with drying wet plant specimens came to the rescue; he dried the papers safely and thus helped to break the Nazi Enigma codes).

Robert Brown's appetite for adventure was rekindled in 1800 when Matthew Flinders persuaded the British Admiralty to send an expedition to explore the New Holland coastline. During the preparations, and just married, Flinders intended to bring his new wife Ann Chappelle on the voyage, defying Navy regulations. His plans were detected, and he was forced to remove her from the ship. Brown, as the expedition naturalist, was to supervise the collection of plants, animals and rock specimens, and was to receive £420 in addition to retaining his commission with the Fencibles. Flinders' ship *HMS Investigator* left port in July 1801. Specimens were to be collected *en route*, and Brown made collections at Madeira and the Cape of Good Hope, and wrote up scientific descriptions of around 6,000 plants, before the crew sighted and named Cape Leeuwin on the southwest coast of New Holland in December. Moored in King George the Third's Sound, Brown, helped by the expedition's gardener Peter Good, collected around 500 plant species new to science, including live specimens intended for the collection at Kew Gardens.

The *Investigator* continued eastward along the south coast. Brown climbed what is now Mount Brown in the Flinders range, and in April, the expedition encountered the French corvette *Géographe*, engaged in a similar mission of scientific exploration. Both commanders, Flinders and Nicolas Baudin, were men of science, and exchanged details of their discoveries, despite believing that their countries were in conflict in the Wars of the French Revolution. In fact, the Treaty of Amiens had concluded a peace two weeks previously, though the peace would last for only a year. Flinders named the site of the meeting 'Encounter Bay'. Progressing east, the *Investigator* party explored Port Philip (now Melbourne) and arrived in Port Jackson (Sydney) on May 9th 1802. Around 700 plant species were now accommodated on board, and the weight of soil was becoming excessive. The plant house on the quarter-deck was halved in size, and the live specimens were planted in the Governor's garden. The ship was in poor condition, and leaking. Sailing north, the expedition explored the coast of Queensland, and somehow reached Timor in Indonesia. The state of the ship forced Flinders to curtail the explorations, but he circumnavigated the continent, and confirmed its status as a single mainland. He raced back to Port Jackson, where *HMS Investigator* was condemned as unfit to sail. Brown and the expedition artist, Ferdinand Bauer, remained in New South Wales while Flinders began the voyage home as a passenger on the 12-gun sloop *HMS Porpoise*, with Brown's best specimens on board, and intending to return to Australia (as Flinders called it) with a new vessel to continue the expedition. It was not to be. Soon after sailing, the *Porpoise* was wrecked on the Great Barrier Reef and the specimens were lost. Flinders heroically navigated the ship's cutter back to Port Jackson, and arranged for the rescue of the marooned crew of the ship. He took command of the schooner *HMS Cumberland* to sail to England, but in December 1803 was forced to put in for repairs at Isle de France (Mauritius). War between France and Britain had erupted again, and he was arrested and detained there for nearly seven years. He was paroled and eventually returned home in 1810. His maps and charts of 'Terra Australis or Australia' and his account of the *Investigator* voyages were published in 1814, the year of his tragic early death from kidney disease, only 40 years old.

After Flinders' departure on what would be his protracted trip home, Robert Brown spent the next two years in Australia, collecting more specimens and studying Aboriginal language and custom. He collected another 1200 plant species, many new to science, before leaving for England on the refitted *HMS Investigator* in May of 1804. A total of nearly 4000 plants had been gathered in mainland Australia, Van Diemen's Land (Tasmania) and Timor, sufficient to occupy Brown's studies for the rest of his career. The zoological collection included birds, insects and a live wombat. Back in England, Brown became librarian and clerk to the Linnean Society while working on writing up the findings of the New Holland expedition. His preliminary publication in 1810, *'Prodromus Florae Novae Hollandiae et Insulae Van-Diemen'*, was received by botanists enthusiastically. Sir Joseph Hooker, the Glasgow-trained botanist and Charles Darwin's closest friend, later considered it to be 'the greatest botanical work that has ever appeared', despite the fact that its 250 poorly-produced copies were unillustrated and unindexed. However in the *Prodromus*, Brown described 187 new plant genera, and around 1000 species in all. He was elected to Fellowship of the Royal Society in 1812.

Although the *Prodromus* was by definition intended to be a preliminary publication, the follow-up books were never published. Instead, Brown published a series of important papers, edited the new edition of *Hortus Kewensis*, the catalogue of plants cultivated at the Royal Botanical Gardens, and became custodian of Joseph Banks' herbarium. He declined offers of professorships in botany at Glasgow, Edinburgh, and the newly-formed University of London, but astutely negotiated with the British Museum to transfer ownership of the Banks herbarium and library, in return for maintaining his personal control and a salary of £200 a year for attendance two days per week.

Brown is most widely known for a discovery he made in 1827 while examining some immersed grains of pollen under a microscope. Within and ejected from the grains, he saw particles, about one micrometer in diameter, moving around constantly and erratically. He initially thought that the particles must be alive, but then saw the same type of random motion in pollen from plants dead for over a century. He said this was a *"very unexpected fact of seeming vitality being retained by these 'molecules' so long after the death of the plant."* He observed the same jittery motion with inorganic particles such as dust suspended in liquid, and tiny chips of window-glass and granite, proving the non-living nature of the phenomenon. He experimented with the effect of heat and burning on the 'molecules' and 'fibrils' liberated from various inorganic and organic materials, writing that *"the substance found to yield these active fibrils in the largest proportion and in the most vivid motion was the mucous coat interposed between the skin and muscles of the haddock, especially after coagulation by heat."* Brown reported this strange finding, and other sometimes spurious, observations in a privately circulated pamphlet in 1828, concluding a year later that *"extremely minute particles of solid matter, whether obtained from organic or inorganic substances, when suspended in pure water...exhibit motions for which I am unable to account..."*. The 'Brownian Motion' was explained by a certain Albert Einstein in 1905. His paper in *Annalen der Physik*, entitled *'On the Movement of Small Particles Suspended in Stationary Liquids Required by the Molecular-Kinetic Theory of Heat'* considered a statistical model of the 'random walk' of Brown's particles constantly impacted by the molecules and atoms of the surrounding fluid. (Brown's early use of the term 'molecules' for the particles from the pollen grains referred to what we now know to be amyloplast and spherosome organelles). Einstein's analysis predicted the average distance travelled by the minute particles as a function of the temperature, molecular size and molecular weight of the suspending fluid, indirectly confirming the existence of atoms and molecules. The predictions were verified experimentally in 1908 by Jean Baptiste Perrin, who was awarded the Nobel Prize in 1926, for confirming the physical reality of molecules.

Robert Brown, still working for the British Museum at the age of 84, died in June 1858, with honours from the Royal Society and almost every scientific society in Europe. He was a freeman of the cities of Glasgow and Edinburgh. A modest and shy man, he had declined to be President of the Royal Society, but eventually accepted the Presidency of the Linnean Society, after repeatedly declining it. He never married. His death prompted an extraordinary meeting of the Linnean Society in July 1858. That meeting proved also to be historic: two revolutionary papers on the evolution of life were presented by Alfred Russel Wallace and, persuaded by Joseph Hooker, by Charles Darwin.

Part Three: The Imperial impulse

23 The Imperial impulse

The growth of the British Empire, and the vigorous expansion of the independent United States of America, offered opportunities for Scotland's engineers, doctors and scientists (a neologism coined in 1833 by the English polymath William Whewell). By the end of the Victorian era, Britain's empire comprised nearly 400 million people and covered 10 million square miles of the surface of the Earth. An industrial revolution was underway, stoking an eager demand for the practical fruits of scientific endeavour. There was need for ships, railways, bridges and telegraphs, as well as diplomats, soldiers and missionaries. Scots would be well represented in all the professional disciplines, and as well as the pull of the Empire and America, there was push. Sheep were proving to be more profitable than people for northern landowners, and the Highland Clearances, combined with potato blight, removed at least 20,000 people from the land. More generally, despite high levels of emigration, Scotland's population grew from around 1.6 million to more than 4 million over the course of the 19th century. In 1800, 17 per cent of people lived in towns. By 1900, it was 50 per cent, and a third of the population lived in the cities of Aberdeen, Dundee, Edinburgh or Glasgow, which with a population of 760,000 was one of the largest cities in the world. 'The second city of the Empire' grew as a result of improved health and longevity, a high birth rate, and migration from the countryside and Ireland. Glasgow's wealth had grown dramatically too, and made her attractive. Facing north-west, Glasgow was hundreds of miles closer to the Americas than any English or Continental port. In the 18th century, the majority of British imports of tobacco and sugar from the slave-powered plantations of the West Indies and the American South came to the city's ports, creating a rich merchant class of so-called 'Tobacco Lords' and handsome Georgian architecture. Exports in return included manufactures like paper, wrought iron, Paisley linen, and 'inkle', an innovative sticky cloth tape that became known as 'Scotch Tape'.

All this made clear the need, and the opportunity, for the further appliance of science. One of the first to foresee this was John Anderson, the Glasgow professor of Natural Philosophy who willed the creation of a new technical university for both men and women. After his death in 1796, his executors used his bequests to implement his detailed instructions for the creation of 'Anderson's University', focused on useful learning *"to the Public for the good of Mankind and the Improvement of Science."* Anderson specified four colleges: Medicine, Law, Theology and Arts - in which he included physics, chemistry, mathematics, and logic. The professor of natural philosophy was to give two courses in physics, the 'mathematical' and the 'ladies course', the latter to be open to both men and women with the objective *"of making the ladies of Glasgow the most accomplished ...in Europe."* The same time period saw the establishment of the *Écoles Polytechnique* in France and the *Technische Hochschulen* in Germany, but for too many years Anderson's Institution was the only technical university in Britain. That would be somewhat remedied with the creation in London of the Royal Polytechnic Institution in 1838. A widening movement for the creation of more polytechnics was promoted by scientists and engineers including John Scott Russell and the Andersonian graduate Lyon Playfair, leading to a Royal Commission recommending on 'Scientific Instruction and the Advancement of Science'. In Scotland, two more institutions were added to the five ancient universities. University College Dundee was founded as a constituent college of the University St Andrews, with the intention of expanding tuition in science and the professions, as opposed to expansion of arts education in St Andrews itself. It would achieve full university status in 1967. The 'School of Arts in Edinburgh', was founded in 1821 by merchant and geologist Leonard Horner who was inspired by Anderson's University. Despite its name, the mission of the School of Arts was to provide practical knowledge in science and technology to Edinburgh's working people. It would become Heriot-Watt University in 1966.

For much of the Victorian period, the causes of science, educational reform and industrial manufacturing were promoted by Albert, Victoria's Prince Consort. The Great Exhibition of 1851, in the Crystal Palace, was one of his major successes and promoted British science and engineering across the Empire and the world. For Scotland, the relative peace and prosperity of the Victorian era fuelled the continuation of the scientific scholarship that had flowered during the Enlightenment

24 The Queen of 19th century science: Mary Somerville (1780-1872)

Mary Somerville was a truly accomplished scientific polymath and writer. Self-educated, she was born in Jedburgh, in the manse of her maternal aunt Martha and her uncle Dr. Thomas Somerville, a Church minister who would be the King's Chaplain in Scotland. Mary was the fifth of seven children of Margaret Charters, daughter of the Solicitor of Customs for Scotland, and a Royal Navy Lieutenant (later Vice-Admiral) Sir William George Fairfax. Three of her siblings died in infancy. Her father was

Illustration 27: Mary Somerville, polymath and science writer - portrait by Thomas Philips

at sea in the American War of Independence when Mary was born, and he spent much of the War as a prisoner of the French. Mary's early years were spent with her elder brother and her mother near the small seaport of Burntisland on the south Fifeshire coast. In her *'Personal Recollections'* she would write: *"During my father's absence, my mother lived with great economy in a house not far from Burntisland which belonged to my grandfather, solely occupied by the care of her family, which consisted of her eldest son Samuel, four or five years old, and myself. One evening while my brother was lying at play on the floor, he called out, 'O, mamma, there's the moon rinnin' awa'. It was the celebrated meteor of 1783."*

Mary was clearly an observant and intelligent child. She was not happy when her mother sent her to learn the 'catechism' of the Church of Scotland, and to attend public examinations in 'edification' (that is, in the building up of faith) in the Kirk. *"This was a severe trial for me; for besides being timid and shy, I had a bad memory, and did not understand one word of the catechism. These meetings were well attended by all...who came to be edified. They were an acute race, and could quote chapter and verse of Scripture as accurately as the minister himself. I remember he said to one of them - 'Peggie, what lightened the world before the sun was made?' After thinking for a minute, she said 'Deed sir, the question is mair curious than edifying'."*

Mary, aged 10, spent a year at Miss Primrose's boarding school in Musselburgh, where she was *"utterly wretched"*. In her early teens at home, she read Shakespeare. On visits to Jedburgh, her uncle, the reverend Dr. Thomas Somerville, a friend of Sir Walter Scott and of the family of the young David Brewster, helped her to learn Latin. She was sent to Edinburgh to the house of her maternal uncle, William Charters, to learn dancing and piano, and to Nasmyth's Academy for ladies, to learn painting. Around this time she secretly studied algebra and Euclid's *'Elements of Geometry'*, defying her father's command not to read mathematics. In a time of political upheaval, she became politically liberal, and somewhat sympathetic to the aims of the ongoing French Revolution, though her father and uncle William *"were as violent Tories as any"*. In Edinburgh, as an increasingly confident, well-connected and attractive young woman, she became fond of attending balls and supper parties. At home, her family were visited by Samuel Grieg, a distant relation from the Charters side of the family. Samuel's father was an officer in the Royal Navy who had been sent by the British government, at the request of the Empress Catherine the Great, to organise the Russian navy. Samuel *"came to the Firth of Forth on board a Russian frigate, and was received by the Fairfaxes at Burntisland with Scotch hospitality, as a cousin."* Mary and Samuel were married in 1804, after he had been appointed Consul to Russia, though they were based in London to avoid the dangerous turbulence in the period after the death of the Russian

Empress. The couple had two sons, Woronzow and William George, born two years apart, and Mary continued to study mathematics, with no encouragement from Samuel, who she said *"had a very low opinion of the capacity of my sex, and had neither knowledge of nor interest in science of any kind."* Samuel Greig died in 1807 aged only 30, from causes unknown. Mary returned to Burntisland, still nursing her youngest son. She resumed her mathematical studies, reading Newton's *'Principia Mathematica'*, Fergusson's *'Astronomy'* and Laplace's *'Mécanique Céleste'*, as well as corresponding with William Wallace, who would succeed John Playfair as professor of mathematics at the University of Edinburgh, and who published a mathematical journal. Of it, Mary noted later: *"I had solved some of the problems contained in it and sent them... as Mr. Wallace sent me his own solutions in return. Mine were sometimes right and sometimes wrong... occasionally we solved the same problem by different methods. At last I succeeded in solving a prize problem! It was a diophantine problem,* [i.e. only integer solutions are allowed] *and I was awarded a silver medal cast on purpose with my name, which pleased me exceedingly."* At the age of 33, Mary was becoming known to the liberal intelligentsia of Edinburgh, including the elderly Professor Playfair and Henry Brougham, who had co-founded the *Edinburgh Review*.

In 1812 Mary married again, to William Somerville, her cousin, and the son of her uncle the Rev. Dr. Thomas Somerville, who held her in great respect and affection. Her Uncle Thomas was delighted. William was an army doctor, interested in science, liberal in politics and encouraging of Mary's work and studies. Not so William's sisters, one of whom wrote to Mary who noted that *"she 'hoped I would give up my foolish manner of life and studies, and make a respectable and useful wife to her brother'. I was extremely indignant. My husband was still more so, and wrote a severe and angry letter to her; none of the family dared to interfere again."* William was appointed head of the Army Medical Department in Scotland and the couple settled at first in Edinburgh, then moved to London, (stopping at Birmingham on their way to visit the 'Soho' steam engine works of Boulton and Watt). With introductions from Professor

Wallace and other Scottish friends, Mary became acquainted with Sir William Herschel, Court Astronomer to King George III and discoverer of the planet Uranus, and his sister Caroline, also a distinguished astronomer and reputedly the first woman to receive a salary as a scientist. Mary, now with two young daughters, continued to make her way in scientific society, attended lectures at the Royal Institution, and visited France, Switzerland and Italy, meeting physicists, astronomers and mathematicians, including François Arago, Jean-Baptiste Biot, Poisson and Laplace. In Florence she was presented to Louise, Countess Albany, widow of the Young Pretender, Charles Edward Stuart, the Bonnie Prince. Mary was unimpressed by the Countess. Back home in Britain, the Somervilles became friendly with William Wollaston, who discovered the elements palladium and rhodium, Thomas Young, who demonstrated the wave nature of light, and Charles Babbage, the inventor of calculating engines.

Mary Somerville's first publication, in the Transactions of the Royal Society (1826), was *'On the magnetizing power of the more refrangible solar rays'*, a report of her observations on the magnetising power of ultraviolet sunlight on steel needles. The paper was read to the Society by her husband William, since women were still excluded from the Society at that time. Her results appeared to confirm the findings of a Professor Morrichini who claimed that exposing steel bars to sunlight resulted in them becoming magnetised. Michael Faraday had examined Morrichini's experiment and concluded that the effect was probably thermal, and not due to true photomagnetism. Somerville herself could not reproduce the effect consistently and it remains a mystery what phenomenon she, and Morrichini, had observed.

A year later Somerville was asked by Henry Brougham (by now a distinguished lawyer and eminent Member of Parliament) to produce in English a condensed edition of Laplace's *'Mécanique Céleste'*, as part of a series of educational books. In its original form the *Mécanique* was five volumes of mathematical analysis applying in detail the universal theory of gravitation, as expounded by Newton, to the motions of all the bodies of the solar system, including the moons of Jupiter, Saturn and Uranus, and comets. Brougham wanted to add a new introduction and explanatory text, and Somerville worked on the project for several years, occasionally consulting Sir John Herschel and Charles Babbage for advice. *The Mechanism of the Heavens'* was published in 1831, covering the first four volumes of the *Mécanique*, and the following year was supplemented by her new introductory text. She said, *"I translated Laplace's work from algebra into common language."* Sales of *The Mechanism of the Heavens* were slow, but it was adopted as a textbook at the University of Cambridge, and brought Somerville more fame. Her breadth of interests, contacts and abilities prompted her to write a second book *'On the Connexion of the Physical Sciences'*, which was published a few years later. It explained recent progress in astronomy, physics, meteorology and geography, and was an immediate best-seller, so much so that updated editions were published annually from 1834 to 1837, and pirated copies circulated in America, and in translation in Germany and Italy. In its third edition, she wrote that anomalies in the orbit of Uranus might point to the existence of an undiscovered planet, which prompted the astronomer John Crouch Adams to embark on calculations that led to the discovery of Neptune. The *'Connexion'* would run to 10 editions and was its publisher's most successful science book until *'The Origin of Species'* in 1859.

In her lifetime Somerville won many international honours. In 1835 she was elected to the Royal Astronomical Society, together with Caroline Herschel, as its first honorary women members. She was invited to an audience with Princess Victoria and was awarded a civil list annual pension of £200, proposed by the Tory prime minister Sir Robert Peel, influenced by Charles Babbage's polemic *'Reflections on the Decline of Science in England'*. Peel wrote directly to Somerville: *"Madam, In advising the Crown in respect to the grant of civil pensions, I have acted equally with a sense of public duty and on the impulse of my own private feelings in recognising among the first claims of Royal favour those which are derived from eminence in science and literature."* In contrast, Henry Brougham was Lord Chancellor in the preceding Whig government, which had recognised several scientists with knighthoods but completely overlooked Mary, much to the disgust of her husband and children. Woronzow was particularly vitriolic about Brougham's oversight. Somerville's pension was increased later to £300 as, despite the relative success of the publication of the *'Connexion'*, the Somervilles met an unexpected financial crisis. William's management of the family finances had long been compromised by a tendency to stretch beyond their means and to provide loans to relatives in need. Without telling Mary, he had guaranteed a loan from Forbes Bank to James Wemyss, son of the minister of Burntisland and cousin to them both. Wemyss ran up more debt, and fled abroad, defaulting on the loan. The Forbes Bank debt would overshadow them for many years. Mary decided that they could live more cheaply abroad, and William's health had declined. The Somervilles moved to Rome for the winter of 1838, never to return to Britain again except for brief visits. Mary continued to correspond and work, not least in order to generate income, and her third book, *'Physical Geography'*, was another masterwork, and the first textbook on the subject in English. It begins *"Physical Geography is a description of the earth, the sea, and the air, with their inhabitants animal and vegetable, of the distribution of these organized beings, and the causes of that distribution.... Man himself is viewed but as a fellow-inhabitant of the globe with other created things, yet influencing them to a certain extent by his actions, and influenced in return."* In it, Somerville anticipated the interconnected view of ecosystems that would emerge in the 20th century.

Mary Somerville continued to live and work in Italy even after William's death in Florence in 1860. Her fourth book, *'On Molecular and Microscopic Science'*, appeared in two volumes, when she was in her 89th year, and was a review of the nature of many plants and chemical substances. Also in that year, she received the Patron's Medal of the Royal Geographical Society for *'Physical Geography'*, her most popular work, and was awarded the Gold Medal of the Geographical Society of Florence. She corresponded with the philosopher and economist John

Stuart Mill to praise him for his book *'Subjugation of Women'*, and when he organised a petition to Parliament to give women the right to vote, he had Somerville make the first signature. She wrote: *"Age has not abated my zeal for the emancipation of my sex from the unreasonable prejudice too prevalent in Great Britain against a scientific and literary education for women."* She died in Naples, peacefully in her sleep, at the age of ninety-one, and is buried in the English Cemetery there. She is remembered widely in the names of *Somerville Island*, British Columbia; *5771 Somerville*, an asteroid in the Main Belt; *Somerville Crater* on the moon; and by the famous *Somerville College*, founded in 1879, seven years after her death, by the University of Oxford to provide non-denominational education for women.

25 Father of modern optics: David Brewster (1781-1868)

Illustration 28: Sir David Brewster with his Stereoscope, by D.J. Pound

Jedburgh, the Scottish borders birthplace of Mary Somerville in 1780, was also the place where David Brewster was born the next year. He was the third child and second son of James, the Rector of the Jedburgh Grammar School, and his wife Margaret Key, who died aged 37 after the birth of their sixth child, when David was only nine years old. David and his three brothers were raised largely by their sister Grisel, and the brothers were educated for careers in the Church of Scotland.

David showed an early interest in science and literature, and was mentored, like Mary, by her uncle, the local minister Dr. Thomas Somerville, in writing and composition. He found another local mentor in the ploughwright and amateur astronomer James Veitch, who helped David construct his first telescope at the age of ten. David left the Grammar School to attend the University of Edinburgh aged 12, corresponding frequently with Veitch to discuss telescopes, electrical machines, and astronomy, and walking the 45 miles home at the end of every term. He graduated with an M.A. from Edinburgh in 1800 and remained in the city to study divinity, as well as working as a tutor and writer for the *Edinburgh*

Magazine', which mixed scientific with literary articles, and of which he would later become the editor. He was ghost-writer of *'The History of Freemasonry'*, published by Alexander Lawrie, though there is no evidence that Brewster was ever a freemason himself. He was licenced to preach by the Presbytery of Edinburgh and gave his first sermon in 1804 in the huge church of St. Cuthberts, known as the West Kirk, to a large congregation. His preaching around Edinburgh and Leith continued successfully for a while, but he was a nervous public speaker and acutely sensitive to criticism, on one occasion even fainting when asked to say grace at a large dinner party.

Acquiring other streams of income became a necessity. Brewster over-optimistically considered applying for the chair of mathematics at the University of Edinburgh, in succession to John Playfair, but abandoned the idea, probably aware that the lecturing duties would have been stressful. The chair was awarded to John Leslie, who was reputed to be 'an extreme Whig and atheist' and whose appointment faced vigorous opposition from the local churches, who accused Leslie of 'heresy' for statements he made on 'cause and effect' in his work *'An Experimental Inquiry concerning the Nature and Propagation of Heat'*. During the 'heated' dispute, an anonymous pamphlet appeared, written by 'A Calm Observer', purporting to take the side of the ministry, but actually comically satirising it and supporting Leslie. The author was Brewster. In the pamphlet, he referred ironically to comments of Professor Playfair, who had opined that clergymen should have no time to interfere in secular issues. Brewster wrote: *"Mr. Playfair's reasoning...can never be extended to the polite metropolis of Scotland. Here there are no sick to visit, no dying to pray for, no children to examine, no ignorant to instruct, no sinners to convert..."*.

Despite his anxiety in public speaking, Brewster made another unsuccessful application for a chair of mathematics, this time at St Andrews. Meanwhile, encouraged by his university friend Henry Brougham, he began serious experimentation in optics, particularly in the diffraction of light. In 1808 he was invited to edit a new publication, the *'Edinburgh Encyclopaedia'*, which provided him with both a basic income and access to a global network of scientists and scholars. Simultaneously Brewster wrote articles for the rival *'Encyclopaedia Britannica'* which at the time was produced and published in Edinburgh. He was elected a member of the Royal Society of Edinburgh and was awarded honorary degrees by the Universities of Aberdeen and Cambridge. Visiting eminent scientists in London, Cambridge and Manchester, and while travelling, he *"invented between Woodford and Epping the Katadioptric telescope"*, a design that uses both lenses and mirrors to reduce the size of the instrument and provide superior image correction. With his new-found financial stability, he began to visit the family of Mr. James 'Ossian' Macpherson of Portobello, and more especially his four daughters. He married the youngest, Juliet, four years his senior, and honeymooned in the Trossachs, accompanied by the bridesmaid (Juliet's sister Anne) and the minister who performed the ceremony. Brewster's diary of 31st July 1810 noted *"MARRIED. Set off for the Trosachs"* in the midst of entries such as *"Thought of new theory of the sun"*, *"Invented new method of measuring crystals..."* and *"Read paper before the Royal Society"*.

Throughout his life Brewster claimed, probably correctly, that he had independently invented the lightweight annular lens, now known as a 'Fresnel' lens, in 1812. He was certainly instrumental in later life in persuading the Northern Lighthouse Board to adopt these lenses. In 1813 Brewster published his first major work, *'A Treatise on New Philosophical Instruments'* which described some of his innovative optical tools and experiments. He had been investigating the effects of transmission, reflection and refraction on the polarization of light. He discovered the relationship - now known as Brewster's Law - that the maximum polarization of reflected light is achieved when unpolarized light falls on the surface of a transparent medium so that the refracted rays make an angle of 90° with the reflected rays. The special angle of incidence to the surface at which this occurs is known as the Brewster Angle. Similarly, if light polarized in the perpendicular plane is incident on a transparent surface at the Brewster Angle, it will be perfectly transmitted through the surface, with no reflection. The exact value of the Brewster Angle (also called the 'polarizing angle') depends on the ratio of the refractive indices on either side of the reflecting surface (specifically, by the cotangent of the ratios). For a glass surface in air, the Brewster Angle is around 56°, depending on the characteristics of the glass. Brewster reported these findings in a paper read to the Royal Society of London in January 1814, and it had immediate applications in devising efficient polarizers, by using stacks of glass plates at the Brewster Angle to the incident light. Today, Brewster's Law has many routine applications in modern optics, in photography, microscopy, and laser design.

Brewster had made another discovery. He already knew that most transparent crystalline materials have a single 'optical axis'; that is, a single direction through the crystal where transmitted light experiences no *birefringence*. If unpolarized light travels through the crystal in other 'off-axis' directions, it experiences two different refractions, one for each of its orthogonal polarizations, resulting in a double image. In studying the structure of the mineral crystals mica and topaz, Brewster found that they had not one, but two optical axes, the first discovery of a 'biaxial' crystal. Within a few years he had studied and classified hundreds of crystals by their optical properties, discovered the effects of heat and compression on their birefringence, and deduced the general laws of '*photoelasticity*', which created the modern sciences of polarimetry and optical mineralogy. The birefringence of calcite crystals and knowledge of Brewster's Law would soon be used by William Nicol of Edinburgh in his ingenious invention of the 'Nicol prism' that became the standard tool for polarizing light for a century and a half.

In July 1814 Brewster embarked on his first tour of the Continent, reluctantly leaving Juliet behind with their two infant sons, James and Charles. Travelling via England where he met William Herschel, in Paris he met French scientific leaders including Arago, Biot, Poisson and Laplace, and was received warmly and openly despite the ongoing

Napoleonic Wars. The wars had hampered exchange of information between French and British scientists in the past - Arago and Biot in particular had made studies of polarized light about which Brewster was unaware. Biot agreed to write the article on 'magnetism' for the *Edinburgh Encyclopaedia*. Brewster was 33, with unassuming manners, brown curly hair and a very youthful appearance, so that his French hosts are reported to have exclaimed, *"What! Is that boy the great Brewster?"* On his return to Scotland to resume his editorial duties, he was awarded a fellowship of the Royal Society of London, and then its Copley, Rumford and Royal medals, as he continued to experiment and publish many papers on optical instruments and the polarization of light. The *Edinburgh Encyclopedia* appeared volume-by-volume over a period of 12 years, during which Brewster and his wife had two more sons and a daughter, Margaret Maria Gordon, who would be his biographer. She recorded that the family were visited by sudden tragedy in 1828 when Charles, aged 15, drowned while swimming in the River Tweed near Allerly, their home near Melrose. Brewster had been haunted by a fear of drowning all his life, and was shocked and distraught. A year later his near-neighbour, Sir Walter Scott, wrote, *"Dr. and Mrs. Brewster are rather getting over their heavy loss, but it is still too visible on their brows, and that broad river lying daily before them is a sad remembrancer."*

Like Charles Babbage, Brewster and his old friend Henry Brougham had become concerned by the stuffiness of the Royal Society of London and what they saw as the decline of science in Great Britain (though Babbage's work referred carelessly to the 'Decline of Science in England'). Brewster proposed in response a 'British Association for the Advancement of Science' and they co-convened its first meeting in York, in September 1831. The meeting lasted for a week, and included presentations of academic papers, and popular lectures and exhibitions. The formula was a success, and since then, the 'British Science Association' meetings have taken place annually almost without interruption.

As an admirer of the work of Isaac Newton and his corpuscular theory of light, Brewster researched and published a short biography, *Life of Sir Isaac Newton* in 1831. But he was again in financial difficulty, and needed income. He considered and rejected an offer of ordination in the Church of Ireland, and applied for the chair of natural philosophy at Edinburgh, vacated on the death of Sir John Leslie. His competitor for the job was James David Forbes, a young man Brewster had mentored and received several times at home at Allerly. Forbes was ultimately chosen for the post, casting Brewster into gloom and severing their friendship for several years. Brewster continued work on his *Treatise on Optics*, a summary of much of his work so far, which appeared in 1838. In the same year he was appointed to a post at the University of St Andrews, as Principal of the United College of St Salvator and St Leonard, a position arranged for him by Lord Brougham, who had already intervened by securing Brewster a knighthood and a government pension of £300 a year. Brewster moved to St Andrews with his wife and daughter, was able to buy the old house called 'St Leonard's', and attempted to invigorate science at a university more noted for its faculties of divinity, moral philosophy and humanities. He met with resistance and an attempt to remove him from his post on the grounds that he was a religious dissenter who supported the breakaway Free Church of Scotland. Brewster was indeed an elder of the Free Church, but not a minister, and the attempted coup failed. At St Andrews, he delivered lectures which were said to be *"perfect models of simple, felicitous exposition of the laws of optics and mineralogy"*. He invited his friend Henry Fox Talbot to demonstrate his new 'calotype' photographic process, for the first time, and was influential in promoting the early advances in photography. He published *The Martyrs of Science* a biography of Galileo, Tycho Brahe, and Kepler, describing their discoveries and persecutions at the hands of the (Roman) church.

Lady Brewster was in declining health and despite visits to the best doctors, Juliet died in January 1850, at the age of 74, and after nearly 40 years of marriage. She was buried at the ruins of Melrose Abbey, beside her son 'Charlie'. In April, Sir David took time off and travelled with his daughter Margaret to Brussels, Antwerp and Paris where they met his old friend Arago, now director of the *Observatoire*, and were visited almost every day by Lord Brougham, coincidentally in Paris at the same time. The next year his youngest son married, but his eldest son James died aged 50, in India. Around this time, Brewster's studies of the Bible, and of the work of Isaac Newton, convinced him that life must exist on many other planets. He published these thoughts in his 1854 book *'More Worlds than One'*, and a year later his studies of Newton produced *'Memoirs of the Life, Writings and Discoveries of Sir Isaac Newton'*.

Despite occasional bouts of ill-health Brewster was an energetic septuagenarian. He continued to travel and publish, and took a lively interest in de-bunking the Victorian fads of spirit-rapping, ghostly visions, and table-turning. Travelling to overwinter in Cannes in November 1856, he met Jane Kirk Purnell, a young English lady on her way to Nice. He followed her there after a couple of months, and by the Spring they were married. She was 30, he was 75, and Margaret wrote that his marriage to 'Jeanie' brought *"a great accession of happiness to his future life; he found in her a most attached and appreciating companion during the years of brilliant life, social and scientific, which yet remained."* They would have a cherished daughter, Constance, born in 1861. By then Brewster had been awarded the *Legion d'Honneur* by Napoleon III, and had somewhat reluctantly resigned from St Andrews to become Principal and Vice-Chancellor of the University of Edinburgh, with Brougham as Chancellor.

Despite his recognition in the academic community, for much of his life Brewster was unknown to the general public. That changed when he invented two new optical devices: the lenticular stereoscope, and the kaleidoscope. The earliest stereoscopes, used to present 3D images from two-dimensional cards and photographs, were clumsy bench-mounted instruments. Brewster used lenses to unite the left-hand and right-hand stereo images, which allowed a reduction in size. Hand-held 'Brewster Stereoscopes' were manufactured, and one was admired by

Queen Victoria when Brewster demonstrated it at the Great Exhibition of 1851. The kaleidoscope, from the Greek word for 'beauty', was based on a number of mirrors tilted at angles to fragment the image of a rotatable coloured object. Brewster had experimented and played with designs as early as 1814, and his friend Sir George Mackenzie predicted its popularity as a toy. The prediction proved correct, but Brewster's kaleidoscope had been demonstrated before he made his patent application, and hundreds of thousands of poor copies were produced and sold within three months, so that he profited little from his invention. To explain the correct designs Brewster published *'A Treatise on the Kaleidoscope'* and wrote to his first wife, *"...gentlemen assured me that had I managed my patent rightly, I would have made one hundred thousand pounds by it!...the mortification is very great."*

Sir David Brewster died at Allerly, aged 86, still working and until the end an energetic Principal of Edinburgh University. He had published around 300 scientific papers and many books, articles and reviews. He was a life-long Christian, liberal and pacifist, and was President of the International Peace Congress held in London to coincide with the Great Exhibition of 1851. His contemporary and sometime rival William Whewell called Brewster *"the father of modern experimental optics"*. He was buried at Melrose Abbey beside Juliet and their son Charlie.

26 Pioneering polarizer: William Nicol (c.1770-1851)

Nothing much is known of William Nicol's education or childhood, and even the date of his birth is a bit uncertain. His parents were John Nicoll and Marion Fowler, and William was born at Humbie, in East Lothian, between 1766 and 1770. His interest in science was stimulated as a teenager by his uncle, Henry Moyes, who was a talented public lecturer in chemistry and natural philosophy. Moyes was blinded by smallpox in infancy, but had excelled at the Kirkcaldy grammar school and met Adam

Illustration 29: Henry Moyes and William Nicol after portrait by J. Reubens Smith

Smith on one of Smith's strolling evening reveries. Smith advised on the boy's studies and later sent Moyes to David Hume in Edinburgh with letters of introduction. Hume arranged for a bursary, and Moyes completed medical studies in Edinburgh and Glasgow. Finding that his blindness was an insurmountable obstacle to the practice of medicine, Moyes became an admired lecturer, always helped by an assistant, first in Edinburgh, then throughout England, and then including lecture tours to Ireland and the cities of the newly-formed United States of America. Around the age of 15, William Nicol began assisting in his uncle's demonstrations of chemistry and optics, travelling extensively around the north of England. In 1807 Moyes died suddenly at the age of 57, in the middle of a lecture course in Doncaster. Nicol was already around 37 years old. He acquired his uncle's equipment and lecture notes, and for many years continued to deliver lectures on a circuit around the English provincial towns.

Nicol had a keen interest in geology and developed great skill as a lapidary, becoming adept at preparing thin slices of minerals and fossils by cementing them to a glass microscope plate and polishing them down to expose their fine structure. He began experimenting with Iceland Spar - transparent crystals of calcite (calcium carbonate) - which has the interesting property of doubling the image of anything placed underneath. Christiaan Huygens and David Brewster (who knew Nicol) had realised what was going on. The crystals were exhibiting double refraction, with light of different polarizations taking different paths through the material - the property of 'birefringence'. Since ordinary unpolarized light consists of a mix of both 'vertical' and 'horizontal' polarizations, two images were the result for any light which travelled through the crystal in any direction except along the special 'optical axis' where no birefringence occurred. Light travelling off-axis through the crystal was split into two different polarizations called the ordinary (o) and extraordinary (e) rays. Nicol also knew from Brewster's work that if plane polarized light travelling in an optically dense material is incident on an external surface at an angle greater than a certain critical angle, then it is totally internally reflected within the material. Ingeniously, Nicol combined these facts to design the world's first light polarizing prism. In principle, a single block of Iceland Spar could be used as a polarizer if it was long enough for the e-ray to be sufficiently displaced from the o-ray so that one of them could be masked off. In 1828, Nicol made a practical prism by sawing a block of Spar in half, polishing the faces, and gluing the halves back together with a cement - Canada Balsam- with a refractive index lower than that which the o-ray experiences in the crystal. He chose the dimensions of the prism and the angle of the sawcut so that the o-ray was incident on the glue

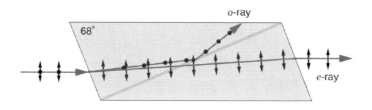

Illustration 30: The principle of the Nicol Prism

interface at greater than the critical angle, and so was totally internally reflected and diverted out of the top face of the prism, leaving the polarized e-ray to pass through in the original direction. He published his invention the next year (though it would probably have been patentable) in a two-page paper in the Edinburgh New Philosophical Journal, in which he described the prism's construction in detail but rather opaquely, given the lack of a diagram. Nicol's short paper was largely ignored outside Scotland for several years until an anonymous German physicist published a paper in *Annalen der Physik und Chemie* in which he praised the performance of Nicol prisms in glowing terms. The German paper was read by Henry Fox Talbot (the pioneer of photography and also a close friend of Brewster) who confirmed the beauty of Nicol's invention. Fox Talbot wrote in the *London and Edinburgh Philosophical Journal* that *"Having read this testimony to the merits of the instrument, I was desirous of making a trial of it, and I caused one to be constructed according to the directions given by the inventor. I found the performance of it fully justified the praises of the writer in the foreign journal. It may be described in a few words by saying that it possesses the powerful polarizing energy of the tourmaline united with perfect whiteness and transparency..."* Nicol prisms (often just referred to as 'Nicols') quickly became the standard tool internationally for polarizing light, or for analysing polarized light, for more than a century, and polarizing microscopes using Nicol prisms became important for the study of mineral and fossil samples. One minor, but interesting use of Nicol prisms was by German physicists exploring the strange phenomenon where the polarization state of light can be fleetingly detected by the human naked eye, through the transient appearance of 'yellow tufts' or 'brushes' ('gelben Büschel') in the vision. These fleeting images, now called in English 'Heidinger Brushes', seem to be related to the structure of our retinal cells, which makes their response to some extent sensitive to the polarization state of incident light.

Nicol retired from his peripatetic lecturing tours in the early 1830s and spent the rest of his years living and studying quietly in Edinburgh, in a house shared with his unmarried sister Marion and two housekeepers. In 1847, a young man named James Clerk Maxwell visited the elderly Nicol in his Edinburgh laboratory. The old man gave him two of the precious prisms, perhaps inspiring Maxwell in his studies which would revolutionise our understanding of the nature of light. Nicol died at home in 1851. His prisms would for many decades be a vital tool in optics, material science, astronomy, and microscopy until superseded by cheaper alternatives, and ultimately the invention of 'Polaroid' polarizing film in 1930 by Edwin Land.

27 The Stirling Engine: Robert Stirling (1790-1878)

The science of thermodynamics is the study of the transformation of energy - including heat energy. The laws of thermodynamics state that energy cannot be created or destroyed, merely transformed; and that in any energy transformation in a closed system, some energy is lost in the form of waste heat, or entropy, thus making perpetual motion machines impossible. This places a theoretical maximum on the efficiency of any engine, for example one which transforms heat into motion. The maximum theoretical efficiency of an idealised 'heat engine' was deduced in 1824 by the French physicist and military scientist, Nicolas Sadi Carnot, to depend only on the difference in the temperatures of its heat 'source' and heat 'sink'. The real engines which most nearly approach the maximum possible efficiency are known as *Stirling Engines*, named after Robert Stirling, a Scottish clergyman and amateur engineer.

Robert Stirling was born the third of eight children near Methvin in Perthshire. His grandfather, Michael Stirling of Dunblane, had invented a water-powered threshing mill, which processed all the corn on his farm, and Robert's father Patrick helped in the maintenance of the rotary threshing and riddling machines. Robert attended the University of Edinburgh aged 15, followed by his younger brother James, and for three years studied Latin, Greek, logic, law, and under Professor John Leslie, mathematics. He then enrolled at Glasgow University to study divinity, before completing his final year of divinity studies, back at Edinburgh, in 1815. He was licenced to preach by the Church of Scotland and began his ministerial career at Laigh near Kilmarnock.

The year of 1816 was known infamously as the year without a summer, now known to have been due to the previous year's violent eruption of Mount Tambora in the Dutch East Indies. This was the largest volcanic eruption in recorded human history. Atmospheric ash and SO_2 reduced average temperatures globally by as much as 3°C and caused widespread crop failures worldwide. In this cold year Stirling appropriately filed a patent for a heat exchanger he called an 'Economiser'. The basic version comprised just two parallel fluid channels in contact through a metal plate. Hot fluid (liquid or gas) passing down one channel was cooled, while cool fluid passing in the opposite direction in the other channel was heated. This is the simple general principle of modern heat exchangers, and it was never patented before Stirling. He then described a practical application of his design: an improved 'furnace' equipped with two flues, to enable heat recovery. Air was drawn into the furnace in the first flue, and the hot combustion gases expelled through the second flue, heating it up. The actions of the flues are then reversed; air is drawn in through the heated second flue and exhausted through the colder first flue. This improves the efficiency of the furnace because some of the heat of the exhaust gases is used to warm the air fed into it - the principle of exhaust heat recovery. Finally, Stirling's patent described the application of these ideas to the design of a fully-functioning engine. This ingeniously used a cylinder, externally heated at its top end and containing two pistons. The first piston (called the 'plunger' or 'displacer') was actually a sealed hollow vessel designed to move freely up and down within the main cylinder, displacing air to and fro between the hot and cold ends of the engine. In the first part of the cycle the increased pressure of heated air moves it past the plunger and into contact with the working piston, driving it down and powering a flywheel via a crank. The air cools and loses pressure at the cold end of the main cylinder helping to allow the working piston back up again, before the air is displaced by the plunger back to the hot end ready for the cycle to start again. Crucially the plunger was also designed to act as a heat exchanger, one of the reasons that the efficiency of the engine was so high. The hot air passing the plunger/heat exchanger in the power stroke reheated the plunger, storing heat energy ready to be transferred back to the used, cool air in the next part of the cycle. The used air therefore arrives at the hot end needing less energy than if it was drawn in cold from outside. Also, the heated air (the 'working fluid') was sealed in the cylinder, continually recycled and never exhausted to the atmosphere, which would have wasted heat. In Stirling's original design of the engine, the heat exchanger/plunger consisted of an airtight hollow vessel made of thin sheet steel and brass. Stirling worked with Thomas Morton, a Kilmarnock wheel-wright and turner, to construct a working two-horsepower engine which was used as a water pump at an Ayrshire quarry.

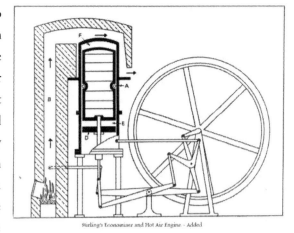

Illustration 31: Schematic of an early Stirling Engine from patent of 1816

Robert's brother James had become a civil engineer, and between them they designed a variant engine configuration with separate interconnected cylinders for the displacer and working pistons, filing another patent application in 1827. They also made the engine 'double acting', that is, producing power on both the upstroke and downstroke. This they achieved by using two displacers, one to act on the topside of the power piston and the other below it, and they increased the power output of the engine by using pressurised air as the working fluid. By 1840 they had improved the design of the displacer/heat exchanger by making it porous and introducing materials such as glass strips to channel the air more effectively. They filed another patent and built two working engines, the first of 21 horsepower and the second producing 45 horsepower. These were used to drive the machinery at James's

Dundee Foundry Company, and the practice and principles were presented by James to the Institute of Civil Engineers in 1845. The engines eventually suffered fracture of the air vessels due to high temperature failures of the cast iron cylinders, but the principle had been well proven. Stirling donated two models of an engine to each of the universities of Glasgow and Edinburgh. The Glasgow model was used for many years by Lord Kelvin to demonstrate to his students some of the principles of thermodynamics.

Robert Stirling continued to combine his engineering ingenuity with his clerical duties. He was awarded the honorary degree of Doctor of Divinity by the University of St Andrews in 1840. His clerical career faltered when he was severely censured in 1842 by the General Assembly of the Church of Scotland for supporting the suspended ministers of Strathbogie, who opposed the appointment of ministers by the 'lay patronage' of wealthy landowners, an issue which would lead to the breakaway of the Free Church of Scotland in 1843. Stirling was suspended until March of that year but retained his position, and ministered to his congregation during a distressing outbreak of cholera in 1848-49, and for many years thereafter. His inventions made him little money, and he died aged 88 in 1878, leaving behind his wife Jane, two daughters and five sons.

In recent times Stirling engines have attracted new attention because of their high efficiency and their ability to use essentially any source of external heat to drive them, including solar heat. They scale well to small sizes and are used in some commercial domestic Combined Heat and Power (CHP) units to generate electricity. They have been used to power submarines, using compressed oxygen as a long-lasting fuel. Stirling engines can also be used as coolers by driving them in reverse. Supplying mechanical power instead of heat reverses the flow of energy and creates a temperature difference between the ends of the cylinder. The 'expansion' end of the cylinder (which would be the heated end when used as an engine) in this mode becomes cooled, and its heat is expelled out into the environment at the 'compression' end. Single-stage

Stirling coolers can produce temperatures as low as 40 degrees Kelvin. The first long-life active cooling system operated successfully in space was a cluster of four Stirling cryocoolers developed by Philips and used in a NASA mission to cool two gamma-ray detectors. Stirling Cycle heat-pumps can be used in a similar way to extract ambient heat from the outside environment and transfer it to warm the interior spaces of buildings. However at the ambient temperatures typically found in rooms, machines based on another thermodynamic cycle are more economical and powerful. That cycle is the Rankine Cycle, using additionally the evaporation and condensation of a suitable working fluid, and based on the work of another Scottish engineer, William John Macquorn Rankine. But that, as they say, is another story.

28 Anatomy at the edge of legality: Robert Knox (1791-1862)

In Britain in the early 19th century, the growth of knowledge of the anatomy of the human body, and the training of young doctors, depended heavily on the surgical dissection of corpses. And the increasing success of scientific medicine and surgery was increasing the demand for qualified doctors, making the medical profession attractive, but also straining the resources of medical schools to cater for students. Medicine was a growth industry in Edinburgh in particular. It was estimated that three-quarters of the University's students were medics. The number graduating increased

Illustration 32: Robert Knox, anatomist and surgeon

from around fifty per year in the first decade of the century (ten times more than Cambridge and Oxford) to around a hundred per year after 1820. By far the largest medical school in Britain, Edinburgh did not require professions of the Anglican faith, nor onerous entrance exams (though graduation did require the production of a scholarly thesis in good Latin), and it was cheaper than London's purely private schools of medicine. However, by common consensus, and including the derisory opinion of one student called Charles Darwin, anatomy at Edinburgh was being taught badly by Professor Alexander Monro *tertius*, who often just read out the old lecture notes of his esteemed grandfather, and whose dissection classes were uninspiring to say the least. A number of 'extra-mural' anatomy teachers became established, who provided additional and more specialised teaching of anatomy - at an extra cost to students of course. Edinburgh soon boasted the highest-regarded schools of anatomy and the best-trained doctors in Europe. English doctors were sniffy and protective in response, and required Edinburgh graduates to be re-licensed or serve an apprenticeship before they could practice in England because a 'Scotch physician so easily gets the degree of Doctor', when in England the title was reserved for a scholarly few. In reality the Scottish training equipped students far better for medical practice, and many became skilled surgeons. The success of Scotland's universities in medical training meant that dissection classes became overcrowded, and the supply of sufficient numbers of cadavers was a problem. The Murder Act of 1752 had ruled that only the bodies of executed murderers could be used for dissection, and demand outstripped supply, so the prices that anatomists were willing to pay for cadavers inevitably increased. Grave-robbing became such a problem that 'mortsafes' (iron cages set over the graves) became a popular option for the wealthier classes, to protect the bodies of their loved ones from the body-snatchers, or 'resurrectionists'.

In 1827, a man called William Hare and his wife Margaret ran a lodging-house in the busy immigrant district of Edinburgh's West Port. When one of their lodgers died, owing rent, William saw an opportunity. He delivered the body to a local dissecting room at Surgeon's Square, earning himself 7 pounds and 10 shillings. The surgeon who paid him was an up-and-coming anatomist called Robert Knox. Hare considered this to be good business, worthy of expansion. Together with his accomplice, William Burke, they decided to maintain a good supply of bodies to Surgeon's Square, and murdered 16 people over a period of 12 months. The method involved getting their victims drunk, with the help of Margaret and Burke's partner Helen McDougal, and then suffocating them by sitting on their chests while clamping their mouths and nostrils closed – what came to be called 'burking'. The murders continued in the dark wynds and closes of Edinburgh until the autumn of 1828, when an Irishwoman in her forties, Madgy Docherty, arrived from Glasgow to visit her son, Michael Campbell, who had been working near Edinburgh during the harvest season. She learned that Michael had left his lodgings three days earlier. She went to a grocery shop near the West Port to find some food, and was unlucky to meet an Irishman, who said he was a Docherty as well. He offered her a place in his home for a fortnight, to give her time to locate her son. Anyway, she should stay the night to join his Hallowe'en party and have a drink. They returned to his tenement room where his wife Nelly made a meal. 'Nelly' was in fact Helen McDougal and 'Docherty' was William Burke. By next morning Madgy was dead, suffocated by Burke and Hare. Robert Knox's dissecting room assistant was asked to come back to the tenement to inspect "something for the Doctor", that was hidden under a straw bed. By the next day the body was delivered, in a tea chest, to No.10 Surgeon's Square, Knox's dissecting rooms. But a couple of Burke's other lodgers had suspicions and had summoned the police. The body was located at the dissecting rooms and the murder investigation soon began to uncover other missing persons connected to Burke and Hare - pensioner soldiers; old wives; and vulnerable street people. Hare turned King's evidence and made a full declaration to escape prosecution. Burke and McDougal were tried with William and Margaret Hare as chief witnesses against them. On December 25th, Burke was convicted, and to the outrage of the citizenry, the case against 'Nelly' McDougal was found '*not proven*'. William and Maggie Hare were hounded and forced to flee, separately, to England and Ireland. In January 1829, Burke was hanged, and dissected by Alexander Monro *tertius*, who publicly displayed Burke's remains in his dissection room.

Robert Knox was a strong-willed and striking figure who as a young boy had contracted smallpox, which disfigured his face and blinded him in one eye. At Edinburgh's High School he was immensely clever, 'dux' of the School, and he graduated from the University of Edinburgh in 1814 with a thesis titled '*De Viribus Stimulantium et Narcoticorurm in Corpore Sano - The effects of stimulants and narcotics on the healthy body*'. Dissatisfied, like many others, with the anatomy teaching of Alexander Munro *tertius* he took John Barclay's extra-mural classes in the subject and became highly expert. For several years, he served as an assistant surgeon in the British Army, including with the 72nd Highlanders in the 5th Cape Frontier War of 1818-19. He developed strong anti-colonialist views, which may have contributed to his accusation of petty theft against a Boer officer. Unfortunately, the officer was the brother of magistrate and military commander Andries Stockenstrom, who later became Lieutenant-Governor of the Eastern Cape Colony (and ironically, developed his own trenchant anti-colonialist views). One of Stockenstrom's supporters challenged Knox to a duel, who refused, but then Knox was publicly horsewhipped, officially reprimanded, and had his expected promotion cancelled. Whether all this is entirely true or not, Knox had little choice but to return home, and was sufficiently composed on his return voyage to continue his interest in meteorology, making detailed observations early every morning, and he wrote *"Temperature of the North Atlantic Ocean and the Superincumbent Atmosphere"* which was published in the *Edinburgh Philosophical Journal*. He returned to Edinburgh on Army half-pay at the end of 1820. Knox's experiences in South Africa formed his later strange, and highly subjective views on the ethnography of the races of humankind.

The next year, Knox obtained Army permission to go abroad for a year of study and research. In Paris, he worked with the noted geologist, palaeontologist, and comparative anatomist, Georges Cuvier. Returning again to Edinburgh, and remaining on Army half-pay, he married a woman "of lower rank" of whom little is known. They would have seven children together; only two would survive to adulthood. Knox was elected a fellow of the Royal Society of Edinburgh and published an important paper, '*Observations on the Comparative Anatomy of the Eye*' in which he compared the detailed physiology of human and animal eyes, and identified the importance of the ciliary muscles in adapting the focus of human (and avian) eyes. Around this time he persuaded the Royal College of Surgeons of Edinburgh to establish a museum of comparative anatomy, and he became its curator in 1825. He also became part-owner of a school of anatomy, with his former teacher John Barclay, and when Barclay died the next year, inherited sole control of it. Knox proved to be an inspiring and extremely popular teacher. His anatomy class was soon said to be the largest in Britain, with more than 500 students, requiring Knox to lecture three times a day. Each of his lectures was a well-rehearsed performance, and he knew how to put on a show. He compensated for his smallpoxed facial features by dressing in the height of fashion. His student and biographer, Henry Lonsdale said *"Dr Knox was wonderfully got up in the way of costume and was perhaps the only lecturer who ever appeared before an anatomical class in full dress. Being a well-made person his tailoring was all the more effective for his display of the glass of fashion and the mould of form. A dark puce or black coat, a showy vest often richly embroidered with purple across which gold chains hung in festoons, a high cravat, white or in coloured stripes, the folds of which were passed through a diamond ring, a prominent shirt collar, delicately plaited cambrics, watch seals and pendants set off by dark trousers and shining boots completed his outer man. Knox in the highest style of fashion with spotless linen, frill and lace and jewellery redolent of a duchess's boudoir standing in a classroom amid osseous forms, cadavera and decaying mortalities was a sight to behold and one assuredly never to be forgotten."* As well as wittily explaining the mechanics of the human body, Knox encouraged his students to develop *"a desire to know the unknown; a love of the perfect; an aiming of the universal."* By this he meant to inspire a search for the overarching laws and archetypes that defined the anatomy of animals and humans - an approach that became known as 'transcendental anatomy'.

The horror of the 'West Port Murders' of Burke and Hare certainly ruined Knox's reputation and ended his chances of gaining a university position. In the other most notorious delivery from the criminals,

Knox's dissection rooms had received the body of an attractive 18-year-old woman, Mary Paterson. In life, according to her Leith landlady, she was *"much given to drink"* and the night before her death had been out with a fellow lodger in her late twenties, Janet Brown. The previous year, Brown had spent ten days in prison for being 'Drunk, disorderly and creating a crowd'. In his trial confession, Burke said that he *"fell in with the girl Paterson and her companion"* one morning in the Canongate. He and Hare *"disposed of her in the same manner"* as their other victims. Janet Brown was the only near-eye-witness who survived to tell the tale, but she was not called to give evidence at the trial. She was described as 'a girl of the town', meaning that she earned money through prostitution, which was of course disparaged but not illegal. Brown had in fact spoken to the police about the rumoured strange disappearances happening in Edinburgh, and she and Paterson had spent the night in the police station, 'for protection'. Released without charge the next morning, they went to a nearby whisky shop for an early 'budge' - the habit of women in struggling circumstances to have a teacup of spirits, poured from pint jugs, to start the day. There they met Burke who enticed them, with a promise of a proper breakfast, to his brother Constantine's house in Gibbs Close, where they ate bread, eggs, and Finnan haddocks, and drank two bottles of whisky. Mary Paterson fell asleep and Janet Brown left hurriedly when Burke's partner Helen McDougal arrived. Brown never saw her friend again - Mary's body was delivered by Burke and his brother to Knox's anatomy rooms in the early afternoon, earning them £8. Her body was naked, and her skirt and petticoat were later found in the possession of 'Nelly' McDougal. What happened thereafter further disgraced Knox. His assistants, one of whom recognised Mary from previous liaisons, admired "the beautiful symmetry and freshness" of Mary's body, and it was reputedly preserved, maybe appropriately, in alcohol, and then voyeuristically displayed to anatomy students for many weeks. When this was discovered by the rumbustious Edinburgh surgeon, Robert Liston, he was said to have knocked Knox down in fury.

In the aftermath of the Burke and Hare trial, the whole scandalous affair focused public attention on the activities of the Edinburgh dissecting rooms, and on Robert Knox's in particular. His effigy was burnt, his premises were attacked by a mob and its windows were broken. Knox was vilified by the public, but he was not prosecuted, and a committee of the Royal Society of Edinburgh found that he had no case to answer, since he had not dealt personally with Burke or Hare. But he was already unpopular with the Royal College of Surgeons of Edinburgh, and they forced him to resign as curator of the anatomical museum he had founded. Publicly reviled as he was, his lectures were still popular and supported enthusiastically by his students. He continued to publish and teach to large audiences averaging 300 or more. In 1832, when he received a settlement of 100 pounds from the Army in lieu of future pay, he moved to larger premises vacated by the College of Surgeons, and lectured twice a day, with a special lecture on ethnology on Saturdays. However the same year, the Anatomy Act introduced government licencing of anatomists, and allowed them legal access to unclaimed corpses in prisons, workhouses and hospitals. The Act also enabled the donation of bodies by next-of-kin in exchange for burial expenses, but the overall effect was to reduce the number of 'subjects' available to the anatomists. The University of Edinburgh made compulsory its own classes in practical anatomy, and Knox's livelihood was under threat. In the next few years he applied for professorial posts at Edinburgh, in anatomy, pathology and physiology. This was optimistic, given his damaged reputation, and his outspokenly sceptical views on organised religion, the administration of Edinburgh University, and the capabilities of his fellow doctors. Of his three rivals for the chair of physiology, Drs. Thomson, Reid and Carpenter, he wrote: *"Nothing throughout this whole business has surprised me so much as the resemblance between the three candidates; their repeated and extraordinary failures; their bolstered up reputations; their total want of all originality; their unpopularity with the student or the taught; their powers of mystifying the plainest facts..."* Dr.

Thomson got the job, and both of Knox's other applications were opposed and unsuccessful.

Knox's wife died from puerperal fever after the birth of their sixth child in 1842, and his son John died aged four, the next year. Despite his ironic wit and entertaining approach to explaining human anatomy, the popularity of Knox's dissection classes declined. He relocated his class briefly to the Portland Street anatomy school in Glasgow, but found stiff competition in the city from its University and the Andersonian Institution, which caused him problems in acquiring students and cadavers. He placed his children and household into the hands of a nephew, and left for England. Finding that he was almost as unwelcome in London's learned institutions as he was in Edinburgh's, he became a peripatetic public lecturer, published many medical journal papers, and wrote several books. But in a final blow to his prestige, in 1847 the Royal College of Surgeons of Edinburgh found him guilty of falsifying a student's certificate of class attendance, and refused to accept any more certificates from him, ending for good his ability to teach in Scotland. The same year the Royal Society of Edinburgh retrospectively cancelled his election as a fellow, and he was expelled.

Knox's views were strident, notorious and often perplexing. He opposed colonialism, slavery and animal vivisection, and fervently supported the French Revolution and Napoleon. He railed against 'kingcraft' and 'priestcraft' and professed atheism while keeping a Bible by his bed. The subject-matter of Knox's highly opinionated books ranged from the sublime to the ridiculous. In *'Great Artists and Great Anatomists'* he eulogised the lives and work of his hero, the French zoologist and 'comparative anatomist' Georges Cuvier, as well as Leonardo, Michaelangelo, and Raphael. In *'The Races of Men'* he presented crude racial stereotypes and caricatures as evidence, that in human behaviour and anatomy, *"race, or hereditary descent, is everything"*, decrying the characteristics of the 'Hottentots' even though he was a slavery abolitionist and criticised the Boers as *"cruel oppressors of the dark races"*. Of the Celts he was especially critical, describing the *"Furious fanaticism; a love of war and disorder; a hatred for order and patient industry; no accumulative habits; restless, treacherous, uncertain: look at Ireland."* But, he said, the Celt was also *"inventive, imaginative, he leads the fashions all over the civilised world. Most new inventions and discoveries in the arts may be traced to him; they are then appropriated by the Saxon race..."* Knox identified himself as a Saxon and therefore was *"self-confident...of unbounded self-esteem"* and abhorred *"all dynasties, monarchies and bayonet governments, but this latter seems to be the only one suitable for the Celtic man."* Knox's more defensible assertion that speciation *"is the product of external circumstances, acting through millions of years"* attracted one reviewer's accusation of *"bold, disgusting and gratuitous atheism."* Amazingly, despite his unscientific views on race, in 1860 Knox was made an honorary fellow of the Ethnological Society of London.

Knox's final years were spent in the practice of medicine, obstetrics and pathological anatomy at the Free Cancer Hospital in London. He was poorly paid, and lived with his sister, and ultimately, with his only surviving child, Edward. His medical career should have been one of outstanding achievement, but perhaps his prospects were always jeopardised by his sardonic wit, religious scepticism and weird racism. The suspicion that he turned his blind eye to the murderous activities of Burke and Hare was undoubtedly the main cause of his professional disgrace. Knox died in London in the winter of 1862, and was buried in the nonconformist area of Woking's Brookwood Cemetery.

29 Geological time and the antiquity of Man: Charles Lyell (1797-1875)

Illustration 33: Charles Lyell, geologist

During the Scottish Enlightenment, James Hutton made the insightful realisation that the rock formations he observed implied a radical new conclusion: that the Earth was formed over immense ages of time by gradual processes of erosion and sedimentation, combined with huge upheavals of the Earth's crust. Hutton's theories were mostly either ignored or rejected, despite the best efforts of John Playfair to explain them clearly - something Hutton had struggled to achieve. Nevertheless interest in geology grew rapidly. From its formation as a small dining club by 13 enthusiasts in a tavern near Covent Garden in 1807, the membership of the 'Geological Society of London' grew to 400 in a decade and more than 700 by the year 1830. Soon the 'Geological' rivalled the 'Royal' and the 'Linnean' as the pre-eminent scientific society in Britain. Its secretary was a young lawyer called Charles Lyell, who would become the foremost geologist of the 19th century.

Lyell was born, the eldest of ten children, to a wealthy family at Kinnordy House, near Kirriemuir in Forfarshire. His grandfather had made the family fortune by supplying the Royal Navy at Montrose, and his father was a sometime lawyer, translator and botanist who ignited his son's interest in nature. The Lyells took a second country home at Bartley Lodge in Hampshire's New Forest, where Charles spent most of his childhood, educated privately. Aged 18, he went to Exeter College, Oxford, and attended the mineralogy and geology lectures of the Rev. William Buckland, one of the 13 founders of the burgeoning Society of geologists. Lyell graduated in 'classical honours' in 1821 and joined Lincoln's Inn to prepare for a legal career. He was called to the Bar the next year. But his interest in geology was becoming paramount, and after his election as secretary to the Geological Society he spent a summer in Paris, studying and conversing with the French geologists Georges Cuvier and Constant Prévost.

Cuvier was a proponent of 'catastrophism', the view that the Earth has been shaped by a series of violent events such as floods, sudden upheavals and abrupt animal extinctions, a view also propounded by Rev. Buckland with added emphasis on the Biblical scale of the required flooding. Prévost, though one of Cuvier's students, took another view. He was convinced that continuous, gradual and ongoing processes explained the geological features he observed, for example the sedimentary rock strata of the 'Paris Basin'. Lyell was convinced too. Back at the Kinnordy House estate, he studied the sedimentary deposits on the bed of a small lake that had been recently drained, and found that these modern deposits were very similar to the Parisian strata: there was no difference between the ancient and present deposition process. He presented his conclusions to the Geological Society during the winter of 1824, in his first scientific paper, *'On a recent formation of freshwater limestone in Forfarshire'.*

Lyell's father was unimpressed by his son's apparent abandonment of his legal career. Lyell relented and did some legal work on the Western Circuit for a year or so, until his parents' move back to Forfarshire reduced the pressure. In the same year Lyell became a fellow of the Royal Society. Still, additional income was needed, and when the publisher John Murray offered him the chance to write for the Tory *Quarterly Review*, he took it, despite his Whiggish convictions. His articles promoted the case for moderate reforms: he bemoaned the lack of government support for science, and the dominant influence of the Church in higher education in England, in the unreformed ancient universities of Oxford and Cambridge. He seized the opportunity to publicise and point to the work of the Geological Society as a model of evidence-based research. True to this espousal of the value of empirical evidence, Lyell embarked on a fieldwork trip with his friend, the military veteran-turned-geologist, Roderick Murchison, accompanied by Murchison's wife and their Swiss maid. They travelled by coach, initially to study the strata in the Auvergne, before Lyell left the party and travelled on to Italy to study the rock formations around the active volcanoes of Vesuvius and Etna. These studies of ancient rocks and fossils underlying the cones of the volcanoes, coupled with historical records of more recent eruptions, allowed Lyell to compare geological time to human timescales, and would form the basis of the conclusions in his life's major work *'Principles of Geology'*. The first volume was published in 1830. Its theme, as he explained to Murchison, was that *"no causes whatever have...ever acted, but those now acting, and that they never acted with different degrees of energy from that which they now exert."* The first part of the statement was not too controversial, but the second part was. Most geologists, even Prévost, believed that geological activity was more vigorous in Earth's distant, hotter past. Lyell opposed this with the Huttonian view of a constantly, but slowly evolving Earth history, of both the inorganic and organic realms. His central argument that *"the present is the key to the past"* was a concept perhaps inherited from the Scottish Enlightenment statement of David Hume: *"all inferences from experience suppose...that the future will resemble the past."*

The sub-title of *'Principles of Geology'* was *'An attempt to explain the former changes of the earth's surface by reference to causes now in operation'*. Its first volume described the processes of the present world - erosion, sedimentation, crustal elevation, earthquake and volcanic activity - and presented *"a grand new theory of climate"*, which viewed long-term climate changes as a product of changes in the physical geography of the planet. Lyell amassed the evidence for past changes in the climate of both the northern and southern hemispheres, and attributed them to changes in the relative positions of land and sea. In the absence of knowledge of plate tectonics, this was a bold assertion. Nor was he afraid to speculate on the appearance of life on Earth. He agreed with Sir Humphrey Davy's view that the development of the human species was relatively recent, but, strangely, resisted Davy's conclusion that *"in the successive groups of strata, from the oldest to the most recent, there is a progressive development of organic life, from the simplest to the most complicated forms."* Lyell's dogged adherence to 'uniformitarianism' and reluctance to concede any 'directionality' in geological history distanced him not only from most geologists, but also naturalists who were working to establish a theory of the evolution of life based on the fossil record. But a good scientist changes his mind according to the evidence. In the tenth edition of *'Principles'*, Lyell would at last accept the case for evolution and directionality in the fossil records.

In a hectic two years from 1831, Lyall was professor of geology at the newly-founded King's College, London; married Mary Horner (daughter of the Whig reformer and geologist Leonard Horner) in Bonn while on a field trip in Germany; and published the second and third volumes of *'Principles of Geology'*. Volume II mainly concerned *"changes of the organic world now in progress"*. In accordance with Lyell's bias towards slow change over geological timeframes, it included a rejection of the 'incessant mutability of species' as proposed by Jean-Baptiste Lamarck, which also implied the idea that humans could have evolved from 'ourang-outangs' - which Lyell at that time found offensive. In the third volume and his subsequent summary *'Elements of Geology'*, Lyell dealt with the various types of rock strata and their relative (and ancient) ages. Although Lyell was very wary of assigning absolute ages to the strata, and considered the earth to be indefinitely old, his analysis resulted in his division of the Tertiary period into the increasingly ancient Pliocene,

Miocene, and Eocene epochs, and reinforced Hutton's notion of deep geological time and the extreme age of the Earth. The publications were highly popular and became required reading for students and lay people alike. One avid reader of the first volume of *'Principles'* was a young naturalist aboard the HMS *Beagle*, on its second voyage of exploration, to the southern oceans. Charles Darwin became a "zealous disciple" of Lyell's view of gradualistic change over ages of time, and would later write in a letter to Leonard Horner, *"I always feel as if my books came half out of Lyell's brains & that I never acknowledge this sufficiently, nor do I know how I can, without saying so in so many words—for I have always thought that the great merit of the 'Principles', was that it altered the whole tone of one's mind & therefore that when seeing a thing never seen by Lyell, one yet saw it partially through his eyes."* When Darwin returned to England, he and Lyell became friends and correspondents. In inviting Darwin to dinner, Lyell advised him to avoid burdensome duties: *"Don't accept any official scientific place, if you can avoid it, and tell no one that I gave you this advice, as they would all cry out against me as a preacher of anti-patriotic principles."* True to his words, Lyell had already resigned his professorship at King's College. Instead he focused on his fieldwork and writing, served a two-year term as president of the Geological Society, and gave a series of well-received lectures in America. He was knighted in 1848 and became an associate of Prince Albert as well as a government adviser on scientific matters.

By 1856, Lyell was well aware of Darwin's work on the theory of evolution through natural selection. When he read a paper by Alfred Russel Wallace called *'On the Law which has Regulated the Introduction of New Species'* he realised its similarity to Darwin's work and urged him to publish it without delay. Darwin had long been reluctant to publish, and with good reason. Twelve years previously, a book entitled *'Vestiges of the Natural History of Creation'* had been published by an anonymous someone, proposing that the development of the solar system, and life on Earth, had progressed through 'transmutations' - we would say evolution - from simpler forms. It was a best-seller (and read aloud by Prince Albert to Queen Victoria) but highly controversial and attracted outrage from the religiously orthodox. Those outraged included Sir David Brewster and one of Darwin's teachers, the Rev. Adam Sedgwick, professor of geology at Cambridge, who attacked *"the foul book"* in the *Edinburgh Review,* and wrote furiously to Lyell (one of the suspects as author) *"If the book be true...religion is a lie; human law is a mass of folly...morality is moonshine...and man and woman only better beasts!"* The author of *'Vestiges'* was only revealed many years after his death. It was Robert Chambers, born in Peebles, member of the Royal Society of Edinburgh, fellow of the Geological Society, editor of *'Chambers's Edinburgh Journal'* and co-owner of the famous publishing house. Darwin had certainly been an early reader of *'Vestiges'* and was fearful of inciting more clerical hostility. He wrote anxiously to his closest friend, the Glasgow-educated botanist Joseph Dalton Hooker, *"If I publish anything it must be a very thin & little volume, giving a sketch of my views & difficulties; but it is really dreadfully unphilosophical to give a resumé, without exact references... But Lyell seemed to think I might do this...Now what think you?... It will be simply impossible for me to give exact references...my only comfort is, that I truly never dreamed of it, till Lyell suggested it, & seems deliberately to think it adviseable...I am in a peck of troubles & do pray forgive me for troubling you."* Darwin continued to hesitate until June 18th 1858 when he received another paper from Wallace, explicitly describing natural selection. Shocked, the same day he sent it to Lyell as requested by Wallace, who was working in Borneo. Lyell and Hooker immediately arranged for the simultaneous presentation of Wallace's and Darwin's papers to the Linnean Society, which took place on July 1st. Neither author could attend; Darwin was in grief over the death of his baby son from scarlet fever three days before. The papers' conclusions were unsettling even for Lyell's religious beliefs, though he had abandoned the Church of England and effectively become a Unitarian, and he resisted their implications for the creation of humanity.

Darwin was disappointed by Lyell's reservations, but he had succeeded in influencing the thinking of the great geologist. Lyell's last major work, *'The Geological Evidences of the Antiquity of Man'* was published in 1863, four years after Darwin published *'On the Origin of Species by Means of Natural Selection'*. *'The Antiquity of Man'* was a review of the latest geological findings that humans had co-existed with mammoths in the glacial period, which Lyell dubbed the 'Pleistocene', extending human history far into the distant past. Lyell's last work also reviewed Darwin's theory of species evolution, still swinging between approval and scepticism. It was well received and sold widely, challenging "the tacit agreement that mankind should be the sole preserve of theologians and historians." But only in the tenth edition of *'Principles of Geology'* (1867) did Lyell finally accept that progressively more complex forms of life had formed over the course of geological time.

In his last few years Lyell was an active president of the British Association for the Advancement of Science, and in recognition of his immense contributions to geology, he was awarded the highest honour of the Geological Society, its Wollaston Medal. His work had embedded in public consciousness the concept of immense epochs of geological time, and the gradual formation of the Earth's structures through familiar processes still at work today. The last words on his influence can be left to Darwin, who in the *'Origin of Species'* wrote, *"He who can read Sir Charles Lyell's grand work on the Principles of Geology, which the future historian will recognise as having produced a revolution in natural science, yet does not admit how incomprehensibly vast have been the past periods of time, may at once close this volume."*

30 Electric clocks and electric thoughts: Alexander Bain (1810-1877)

Illustration 34: Alexander Bain, inventor

Alexander Bain was one of the most prolific inventors of the 19th century, and is one of the least remembered. Not to be confused with the Scottish empirical philosopher of the same name, this Bain was responsible for startling innovations in the field of electrical instruments, including the invention of the electric clock, an improved electric telegraph, the fax machine, and the fundamentals of television. His inventions made him rich, for a while, and they are all the more remarkable given his unusual educational background. Born near Watten in Caithness, his father was a tenant crofter and Alexander was one of six sons and six daughters. He was not exceptional at school, and left at age 12 to work on his father's farm. His skill was in fixing things, including clocks, and when he was nearly 20 years old he was apprenticed to John Sellar, a clockmaker in Wick. As a new apprentice he attended a public lecture in Thurso entitled 'Light, Heat and the Electric Fluid' which captured his imagination completely. He later said of himself *"When the lecture was over, and the audience were leaving, a few gentlemen accompanied the lecturer, and conversed with him on the subjects of the lecture. There was a humble lad walking behind them, and listening attentively to what was said...he never forgot the lecture, nor the subsequent conversation."* After nearly seven years of clock-making apprenticeship, he left the north of Scotland, briefly for Edinburgh, and then London, 700 miles away from home.

Clerkenwell was the centre of London clock-making, and while working there as a journeyman, Bain went to evening lectures at the new and pioneering Polytechnic Institution (now the University of Westminster) and the National Gallery of Practical Science, also known as the Adelaide Gallery. Both venues had frequent public displays of new inventions, and Bain's imagination was fired again. A confusing range of barely working and impractical instruments were often on display, including clocks that used the properties of static electricity to maintain the swing of the pendulum. Bain thought he could do better. By the middle of 1840 he had made a model of an electromagnetic clock powered by electric current, as well as a 'printing telegraph'. Ingeniously, he also realised that a source of sufficient electrical power could be the ground, if a galvanic action could be set up between two suitable electrodes - plates say of zinc and copper, or zinc and carbon - buried in the earth and connected to the instruments by ordinary wires. He also thought that it was obvious *"to make a common clock transmit its time to other distant clocks...worked by what is termed electro-magnets."* His electric pendulum clock came in two variants. In the first, the pendulum bob is an electromagnet swinging between two permanent magnets, and in the second, the bob has permanent magnets attracted to and fro by electromagnets. In either case, the impulse was maintained by a bar making and breaking electrical contact with the pendulum during its motion. In Bain's 'printing telegraph', the character to be transmitted was selected by stopping a clockwork-driven pointer at the correct location on a labelled disc. In the receiver, the printing type-wheel was rotated into the corresponding position by an electric clock escapement released by an electromagnet, one tooth at a time, by a number of electrical pulses. The appropriate number of pulses was produced by the pointer passing over electrical contacts on the transmitter. The type-wheel in the receiver printed the selected character on a drum of paper.

In both the transmitter and receiver, the pointer on the dial and the printing wheel were continually rotated by their respective clock mechanisms, to increase the speed of character selection, transmission and printing.

Bain seems to have displayed his electric clock model at the Polytechnic Institution, and unsure of the next step, he contacted the assistant editor of *Mechanic's Magazine* for advice. The editor suggested that Bain should meet Charles Wheatstone, who was professor of experimental physics at the new King's College, London. Bain was introduced on August 1st and met again at Wheatstone's home on the 18th, taking his models along. Wheatstone gave Bain £5 for his model of the printing telegraph, with a promise of a further £50 if Wheatstone made a working telegraph unit that could be sold. Bain understood that Wheatstone also ordered two working examples for £150, though that would be disputed. Wheatstone also asked that their discussion should be kept confidential, and work on the electric clock stopped, quoted as saying *"Oh, I shouldn't bother to develop these things any further...there's no future in them."* Bain ignored the advice and continued to work with John Barwise, a chronometer maker in St. Martin's Lane. Presumably unknown to Wheatstone, they applied for a patent on the 10th of October. More than a month later, at the end of November, Wheatstone demonstrated an electric clock to the Royal Society, claiming to have invented it. Bain had delivered the telegraph instruments to Wheatstone the same autumn, but was not paid even to compensate for the cost of materials used. He confronted Wheatstone but after a heated argument, left without his money or his instruments. A bitter feud would ensue, and become public.

The necessary Royal Assignment of Bain's clock patent was delayed by Queen Victoria's first confinement and childbirth; but the patent, number 8783, *'Improvements in the Application of driving power to Clocks and Time Pieces'* was signed on January 8th, 1841. The patent was important, and it was wide-ranging. Apart from the use of switched electromagnets to drive the clock, it covered the use of a central clock to operate a number of other clocks, and a system to distribute time from a master clock to a number of remote 'slave' clocks. With the patent duly signed-off, when Wheatstone attempted in January again to exhibit an electric clock, this time at the Adelaide Gallery, John Barwick succeeded in securing an injunction to prevent him.

The work of both parties continued and in May 1842, Wheatstone did exhibit an electric clock at the library of the Royal Institution, mentioned in the *Literary Gazette* of the 28th as *"Wheatstone's Electro-Magnetic Clock...as it has become familiar to the public."* Bain immediately wrote to the editor to assert his priority, and the result was a furious public exchange of letters for the next two years, played out in the pages of the *Inventor's Advocate* and the *Literary Gazette,* Bain and Wheatstone each claiming precedence in the invention. To Bain's cause were rallied Sir Peter Laurie, chairman of the Union Bank and a previous Lord Mayor of London, and John Finlaison, an Admiralty and Treasury civil servant, Actuary of the National Debt, later the first president of the Institute of Actuaries. Finlaison by a strange coincidence hailed from Thurso, a few miles from Bain's birthplace, and had been impressed by a demonstration of Bain's printing telegraph writing *"On enquiring who was the inventor of this extraordinary apparatus...he was a young man, Alexander Bain... a self-taught genius, from the author's own native spot in the extreme North of Scotland, totally unfriended and hitherto unknown to fame...There was further shown...another of his inventions, consisting of a clock moved by electricity...capable of making any number (say 500) of other clocks simultaneously work together."* Finlaison became a loyal ally.

Wheatstone was already a fierce rival, particularly jealous of Bain's innovations in telegraphy, saying of him, *"He was a working mechanic, who was employed by me between the months of August and December, 1840. Of the true principles of telegraphic communication by electro-magnets, which, aided by the beautiful theory of Ohm, I was the first to determine, he evidently knows nothing."* With his co-author William F Cooke, an army officer, Wheatstone had filed a patent in May, 1837 for an electric 'needle' telegraph where the receiver used needles, moved by electromagnets, to point to letters on a flat board. Experimental electric telegraphs had been demonstrated before, and their 1837 patent was for a clever development of the proposals of Baron Schilling in Germany. Any number of needles could be used depending on the number of characters to be used. A four-needle system was installed on the railway line between Euston and Camden Town in London, and was demonstrated in July 1837. A Cooke-and-Wheatstone five-needle system was installed on the Great Western Railway a year later. Clearly, thoughts and developments of the electric telegraph were quite advanced, compared to those of electric clocks. And the electric clock seems not to have been a part of any written contract or formal transaction between Wheatstone and Bain.

Charles Wheatstone was an odd professor - inventive, but shy and averse to public speaking; acquisitive of ideas, but not really interested in commercial applications; yet prone to becoming entangled in disputes over where credit was due. He argued with Cooke in 1841 over the honour of inventing the needle telegraph, and ludicrously, the dispute went to arbitration with Marc Isambard Brunel supporting Cooke, and Professor J.F. Daniell of King's College in Wheatstone's corner. The result was a draw. Differences were patched up and Cooke and Wheatstone formed, with financiers, the Electric Telegraph Company in 1846, and applied to Parliament for a Bill of Incorporation to allow construction of the wirelines. Alexander Bain's antipathy to Wheatstone was alive and well. He immediately opposed the Bill on grounds of infringement of his 1843 patent *'Certain Improvements in Producing and Regulating Electric Currents, and Improvements in Electric Time- Pieces, and in Electric Printing and Signal Telegraphs'*. This was not snappily titled, but it proved valuable. After Bain and Cooke both gave evidence, the House of Lords committee favoured Bain and suggested an out-of-court settlement. Bain withdrew his objection and was paid £7,500 in compensation and £2,500 for the use of his printing telegraph patents. Furthermore, the Electric Telegraph Company would manufacture and market Bain's clocks, and he would receive half the profits. Bain became a very minor shareholder, and Wheatstone responded to the settlement by resigning from the company. The road was clear for electric telegraphy to proliferate. One commentator, the physicist and journalist John Herapath, wrote in July 1846 that *"the electric telegraph... has been not inappropriately termed 'the railway of thought'."*

The same year, Bain filed another patent together with Robert Smith, a lecturer in chemistry from Blackford, Perthshire, which described another invention that would prove lucrative: a silent 'chemical telegraph' system. The message was sent as dots and dashes. The receiver used discs or strips of chemically treated paper, supported by an electrically earthed metal surface below, and a metal stylus to write marks on the paper according to the signals received. Bain later developed the idea further and invented a system of sending entire documents, not just characters, by telegraph. This further invention used a swinging pendulum to scan the document to be sent, which consisted of metal typeface characters, line-by-line as it moved down, making an electrical contact at each point on a character. The remote receiver printed the copy using a similar, synchronised pendulum with a protruding stylus, which left a mark on chemically-treated paper when an electrical current passed from the stylus. In that way every point on the original document could be copied and transferred to a printed point on the paper. This was what we would now call a facsimile (fax) system. The line-by-line 'raster scanning' of the source document anticipated the methods that would be used by John Logie Baird and other pioneers of television to capture, transmit and reconstruct moving images. Bain did not further develop this

Illustration 35: Bain's facsimile machine (1850)

invention, but in 1865, Abbé Caselli started the first commercial facsimile service between Paris and Lyon.

In 1844 Bain married Matilda Bowie, the widowed sister-in-law of his greatest champion, John Finlaison, and moved his business to Edinburgh, then back again to London after four years. He and his wife would have five children to add to Matilda's daughter from her first marriage. While in Edinburgh, Bain won a contract from the Glasgow and Edinburgh Railway to construct a telegraph line along their route, 46 miles long. The price quoted to the railway company was £50 per mile, much less than the £250 per mile Cooke and Wheatstone were charging the Great Western Railway. Finlaison loaned £3,000 to the project and Bain's brother John helped in construction. Bain was mainly interested in proving the capability of his time distribution system over a significant distance, which he did, using a master electric pendulum clock in Edinburgh connected by a single wire (with an earth return path) to a slave clock in Glasgow. Returning to London with his new wife in 1847, he established a new home at Beevor Lodge in Hammersmith, and, now a successful businessman, he was able to afford five domestic servants and a private tutor for his children.

His next target was the growing market for telegraph systems in the United States of America, where the construction of railways was going full steam ahead. Railways needed signalling, and provided convenient rights-of-way. Bain left for America, leaving his family behind. Unfortunately, he was not alone in spotting the opportunity, and the American competition was intense. Samuel Morse had already built a telegraph between Baltimore and Washington D.C. and other players were joining the fray. When Bain applied for a U.S. patent on the chemical telegraph in 1848, Morse successfully opposed it on the grounds of having made a similar patent application himself earlier that year. Bain appealed to the Federal Court which accepted the priority of Bain's British patent and overruled the U.S. Commissioner of Patents. Bain and Smith received their U.S. patent in October 1849, and their chemical telegraph was commercially licensed and used on more than

2,000 miles of railways. Morse came back on the attack and in 1851 obtained an injunction for infringement of his original patent of 1840, which was so wide-ranging that it covered almost every conceivable aspect of communicating by electricity. The case went all the way, reaching the Supreme Court of the United States, in 1853. The court ruled in favour of Morse. As no comfort to Bain, the broadest, unimplemented patent claims of Morse were rejected, and the judgment is seen as a landmark ruling, often cited in patent law where broad ideas and specific implementations need to be distinguished.

Defeated and financially damaged by Morse's injunction, Bain returned to London in 1851, and after the final ruling his American assets would have been devalued and were liquidated. The legal costs of his failed defence in the Supreme Court had been high. In 1853 Bain applied for a discharge in the Court of Bankruptcy, with total debts of $58,000 (around £12,000). He had few remaining assets apart from his patents, which were surrendered. His links were severed with the Electric Telegraph Company (which was later nationalised and merged with the General Post Office). He went on to concentrate on selling electric clocks, while dabbling with further inventions. Matilda died in 1856 and thereafter Bain seems to have spent little time with his family, returning briefly to the USA in 1860, probably to pursue money owing to him from wound-up assets, and with no obvious result. Back home, the clocks were seen as expensive luxuries and did not sell well. By 1872 Bain was pretty much back where he started, working for a watchmaker in Glasgow, repairing clocks for a living. There, one of his customers was the University's William Thomson (later Lord Kelvin) who recognised Bain's genius and his plight. Professor Thomson arranged a grant of £150 from the Royal Society, and successfully petitioned the Gladstone government to award Bain a Civil List pension of £80 per year. Bain died in 1872, cheated of fame and fortune by bad luck and bad choices. In his life, aside from electric clocks and the chemical telegraph, he patented a long list of other inventions, including a fire alarm; a marine depth sounder; a system for recording ships' direction and speed at sea; a device for producing punched tape and a piano for playing the tape remotely; a current regulator for voltaic cells; a drinking fountain tap operated by pressing the receptacle on a lever, and perhaps too fondly, a device for drawing a measure of liquid from a container, similar to a bar optic for spirits. He is buried in Kirkintilloch, and honoured in the names of the BT (British Telecom) buildings in Glasgow and in Thurso: 'Alexander Bain House'.

31 Diffusion and dialysis: Thomas Graham (1805-1869)

Illustration 36: Thomas Graham, chemist -lithograph by R. Hoffmann

Dialysis is defined as the process of separating particles or molecules in solution by exploiting their differing rates of diffusion, as they pass through a semipermeable membrane. In medicine, failure of the kidneys to remove waste products from blood is a fatal condition. Kidney dialysis has become an effective treatment for the huge number of patients suffering from kidney failure. Today, nearly three-quarters of a million Americans have severe kidney disease, most of these requiring dialysis three times a week, for 3-5 hours at a time. In the UK, around 30,000 people require the same weekly regimen of kidney dialysis.

The science of diffusion and the techniques of dialysis were developed by a physical chemist, Thomas Graham, born in Glasgow in 1805. Throughout his studies at the High School, and the University of Glasgow, he resisted the wishes of his father, a prosperous merchant whose brother and uncle were church ministers at Killearn, who wanted him to study divinity and likewise enter the Church of Scotland. Instead Thomas studied science, sometimes surreptitiously. When he entered the University aged only 13, he became interested in the chemistry lessons of Thomas Thomson and in the natural philosophy taught by William Meikleham. In 1824 he did enrol in divinity classes to please his father, but in fact studied physical chemistry instead, and presented a paper to the Glasgow University Chemical Society on the absorption of gases by liquids. He then persuaded his father that the best place to study divinity was the University of Edinburgh, and promptly enrolled there - to study medicine and chemistry. He set up a small laboratory in his lodgings and continued his work on the absorption of gases by liquids, publishing a paper of that name in the *Annals of Philosophy*. His father realised the situation and visited Edinburgh to destroy the lab equipment, and cut off his son's financial support.

Thomas persisted in his studies while working in the Edinburgh University chemical laboratory, and giving some extramural lectures in mathematics. In 1828 he returned to Glasgow, newly elected as a fellow of the Royal Society of Edinburgh, and taught mathematics and chemistry at the Portland Street medical school, while giving evening lectures on chemistry at the Glasgow Mechanics Institution. Portland Street gave him renewed access to a laboratory, where he investigated the action of charcoal on solutions, platinum catalysis, and gaseous diffusion. By submitting a paper on gas diffusion, '*On the tendency of air and the different gases to mutual penetration*', he became a member of the Faculty of Physicians and Surgeons of Glasgow, which then enabled him to make a successful bid to become professor of chemistry at the Andersonian Institution, and ultimately to be reconciled with his father.

His painstakingly careful experiments measuring the diffusion of gases through porous plugs, fine tubes, and small pinholes, showed an important and useful result. In general, heavier gases diffused more slowly. Specifically, for a gas at constant pressure, its rate of diffusion was inversely proportional to the square root of the mass of its particles (i.e. its molecular weight). So for example, if the molecular mass of a gas is 4 times the mass of another, the heavier gas will diffuse through a porous material at half the rate of the other one. This became known as

Graham's Law, first published in 1831 and winning the Keith Medal of the Royal Society of Edinburgh. Thomas's mother had long supported his ambitions in chemistry, and sadly, learned of his honour on her deathbed. Graham's Law has many practical applications, including the separation of isotopes by diffusion, a process important today in the enrichment of uranium to make nuclear fuel.

Graham went on to do more important work in pure chemistry, showing for example that phosphoric acid gives rise to three different compounds and is 'polybasic', that is, able to donate more than one hydrogen ion from each of its molecules. In 1833, Graham published the first part of a series of books called *Elements of Chemistry* which would become widely influential and was translated into German by Nikolaus Otto. Graham's fame grew quickly and in 1836 he was made a fellow of the Royal Society of London, and the following year was awarded its Royal Medal. The same year he was offered and accepted the professorship of chemistry at the newly-formed University College London (UCL). He never lived in Scotland again, even though he inherited his father's country estate in 1842. During his time at the Andersonian Institution, Graham trained and inspired a whole generation of chemists who would invent and revolutionise many industrial processes, becoming what we would now call 'chemical engineers'. He was so respected and loved by his Andersonian students that several of them followed him to work as his assistants in London, including Lyon Playfair (who became professor of chemistry in Manchester and at Edinburgh) and James Young, who would make a fortune in industrial chemistry and the distillation of paraffin oil.

At UCL, Graham took on a heavy lecturing workload while he continued his research on gaseous diffusion, and on 'osmosis', a term introduced years before by the French physiologist René Dutrochet, to describe the transfer of water across a semipermeable barrier from a solution of low concentration to one of higher strength, tending to equalise the solute concentration. Graham expanded and generalised this concept. In his Bakerian Lecture of 1851 he introduced and explained the concept of osmotic force (or pressure), rejecting Dutrochet's idea that capillary action in the apparatus had an effect, but agreeing that what he dubbed 'osmose' must be of importance in living cells. He wrote, *"...chemical osmose appears to be particularly well adapted to take part in animal economy. It is seen that osmose is particularly excited by dilute saline such as animal juices really are, and that the alkaline or acid properties which these fluids always possess is another most favourable condition for their action on the membrane."* He realised that osmosis was *"the conversion of chemical affinity into mechanical power"* and with great prescience asked *"... what is more wanted in the theory of animal functions than a mechanism for obtaining motive power from chemical decomposition as it occurs in the tissues?"*

From about 1842, Graham had been increasingly involved as a government adviser on a wide range of scientific (and not so scientific) issues including the purity of tobacco and coffee, the quality of London's water supply, the casting of metal for guns and even the proposed tactics of Vice-Admiral Thomas Cochrane, Earl of Dundonald, for the prosecution of the Crimean War. From 1851 Graham had also served as the chemical assayer to the Royal Mint. He resigned his chair at UCL, and then succeeded Sir John Herschel as Master of the Mint in 1854, a post he held till his death. During his tenure he made large cost savings, in the minting of gold and silver coins, and by substituting durable bronze for copper in low-value coins.

Fortunately, Graham continued his research in diffusion and osmosis, and made more breakthrough experiments. Using starched ('sized') wet paper as a membrane (septum), he showed that some materials in a mixed solution, such as sugar, could pass through it, while others, such as gum arabic, could not. In his 1861 publication *'Liquid Diffusion Applied to Analysis'* he identified the fundamental difference between mechanical filtration and molecular diffusion. He wrote, *"The sized paper has no power to act as a filter. It is mechanically impenetrable, and denies a passage of the mixed fluid as a whole. Molecules only permeate this septum, and not masses. The molecules also are moved by the force of diffusion..."* The mechanism involved in this diffusion was subtle. The sugar, as what

Graham called a 'crystalloid', could combine with the water in the starched paper and diffuse through the water in that gelatinous barrier. The gum arabic, as what Graham defined as a *'colloid'*, had little affinity with the water in the starch, and was held back. Graham noted *"It may perhaps be allowed to me to apply the convenient term 'dialysis' to a separation by the method of diffusion through a system of gelatinous matter."* He moved on to invent a simple apparatus to aid more experiments - a 'hoop dialyser' - and used it to demonstrate the dialysis of urea (a 'crystalloid') from urine. As well as identifying the mechanism and the usefulness of dialysis, Graham's work was the beginning of colloid chemistry. He was careful to identify materials - such as starch, gum, albumin and tannin - which had low diffusibility, gelatinous hydrates, and are *"held in solution by a most feeble force"* as colloids, and *"their peculiar form of aggregation as 'the colloidal condition of matter'."* In contrast, he defined crystalloids as highly soluble, and quick to diffuse. He began to use graphite as the diffusion medium and defined two other, different modes of gas motion. In 'effusion', the passage of gas particles into a vacuum through a tiny hole, smaller than the average distance travelled by gas particles between collisions, the rate of gas transfer is also determined by Graham's Law. In 'transpiration', vapour or gas is evaporated from the surface of a liquid, and diffuses into the surrounding gas (or air), again as defined by Graham's Law. He recognised this mechanism as being important in the respiratory tracts of insects and other animals.

For his lifetime's work on the laws of gas diffusion, polybasic acids, catalysis, osmosis, colloids and dialysis, as well as his promotion of chemistry within the British Association for the Advancement of Science, Graham was awarded the highest honour of the Royal Society, its Copley Medal, in 1862. His equations governing the laws of diffusion, deduced from experiment, were confirmed by the mathematics of a young Albert Einstein in 1902. Graham's work on dialysis was considered at the time to be one of his lesser achievements. In fact, it has proved to be one of his most beneficial. Subsequent work by Moritz Traube in Germany produced durable semipermeable membranes robust enough to withstand high pressure, and modern dialysers use hollow synthetic fibres to provide the semipermeable barrier.

Thomas Graham died of pneumonia in 1869 at home in London. He never married. He was buried at Glasgow Cathedral, and is honoured by the striking statue in the center of the city at George Square, and the eponymous chemistry department building of the University of Strathclyde, the descendant of the Andersonian Institution where he held his first professorship. He is owed a huge debt of gratitude by the million or so people who receive life-sustaining kidney dialysis every week.

32 Great ships and solitary waves: John Scott Russell (1808-1882)

In 1851, the Australian Royal Mail Steam Company wanted ships that would make the journey from Britain to the Antipodes taking on coal only once, at the Cape of Good Hope. They turned to their chief engineer, the great ship designer of the age, Isambard Kingdom Brunel, creator of the largest ship afloat, the *SS Great Britain,* which measured 322 feet long, and displaced 3,675 tons. In response, Brunel produced a specification for ships displacing between 5,000 and 6,000 tons. The company baulked at what they saw as the over-ambitious design. So

Illustration 37: John Scott Russell, engineer and naval architect

instead, Brunel commissioned another design for two smaller ships from the well-known Scottish naval architect, John Scott Russell, which would still be able to carry enough coal to meet the requirement. Russell owned the Fairbairn ship-building yard at Millwall which was respected and reliable. Under the contract, two big iron-hulled mail steamers, the *Victoria* and the *Adelaide*, each carrying 200 passengers, were designed, constructed and launched successfully. The *Victoria* won a prize for the fastest passage to Australia: 60 days, including a two-day stay at St. Vincent. By the spring of 1852, Brunel was wondering how to design a ship capable of sailing to the Far East, and back again, carrying all its own fuel. He began discussing plans with Russell for a truly enormous ship, over 600 feet long, and displacing more than 21,000 tons. It would become the *SS Great Eastern*.

John Scott Russell was a superb naval architect and engineer. Among his many achievements, two stand out: first, the construction of the *SS Great Eastern* in collaboration with Brunel; second, the discovery of a strange wave phenomenon which has had many profound implications in the modern fields of fluid dynamics and optical telecommunications. Scott Russell was born the only son of a clergyman, Rev. David Russell, and Agnes Scott in Parkhead, now a part of Glasgow but then a local village. At the age of 12, he studied for a year at the University of St Andrews before enrolling at Glasgow University to study mathematics and natural philosophy, graduating when he was only 17 years old. He moved to Edinburgh where he founded a preparatory school for the University called the South Academy. He was encouraged by his old professor of geometry at St Andrews to teach mathematics and science at the Leith Mechanics Institute, and at Edinburgh University, where his lectures were highly popular and so well-attended that *"when he commenced his second course of lectures, the classrooms of his former master and actual rival were rapidly emptied."* On the vacancy of the chair of natural philosophy at Edinburgh, due to the death of Sir John Leslie in 1832, Scott Russell (then aged 24) was temporarily elected to the post, pending the appointment of a permanent new professor. He was encouraged and invited to apply, but declined to compete with another candidate he greatly admired, David Brewster, who ironically did not get the job but later became Principal of St Andrews University. Scott Russell was in any case more interested in industrial applications, and moved away from academia. He briefly ran the *Scottish Steam Carriage Company*, which offered steam-car passenger transport between George Square in Glasgow and the Tontine Hotel in Paisley, until a fatal accident ended the service. It was while consulting for a company operating a passenger steam-boat service on the Edinburgh and Glasgow Union Canal, that John Scott Russell discovered a most amazing phenomenon.

He was conducting experiments at the Union Canal near Hermiston to determine the most efficient design for canal boats, observing a boat being towed along rapidly by horses. When the boat stopped, the bow wave continued forward in a very unusual way. He described his discovery like this: *"I was observing the motion of a boat which was rapidly drawn along a narrow channel by a pair of horses, when the boat suddenly stopped—not so the mass of water in the channel which it had put in motion; it accumulated round the prow of the vessel in a state of violent agitation, then suddenly leaving it behind, rolled forward with great velocity, assuming the form of a large solitary elevation, a rounded, smooth and well-defined heap of water, which continued its course along the channel apparently without change of form or diminution of speed. I followed it on horseback, and overtook it still rolling on at a rate of some eight or nine miles an hour, preserving its original figure some thirty feet long and a foot to a foot and a half in height. Its height gradually diminished, and after a chase of one or two miles I lost it in the windings of the channel. Such, in the month of August 1834, was my first chance interview with that singular and beautiful phenomenon which I have called the Wave of Translation, a name which it now generally bears..."*

Scott Russell knew that in his 'Wave of Translation' he had observed something fundamentally important, and he built an experimental tank in his garden to continue his studies of it. He developed his observations of the strange solitary waves and measured their key properties: stability and resistance to dispersal, so that they can travel over great distances; their high speed; and the dependence of speed on the width and depth of the water channel. The wider relevance of Russell's solitary waves only became clear in the 1960s when scientists began to use digital computers to study non-linear wave propagation in fluids and in solid materials. It became obvious that many phenomena in physics, electronics and biology can be described by the theory of 'Russell's Solitary Waves' as they are now known in fluid dynamics, and of 'solitons' as they are called in the field of fibre-optics.

Solitons have caused especially great excitement in the fibre-optic communications industry. The qualities of solitary waves which intrigued Russell - the fact that they do not fragment, disperse, or lose strength over distance - prompted Akira Hasegawa at AT&T Bell Labs to propose in 1973 that solitons could exist in optical fibres if a sufficiently powerful light pulse exploited the non-linearity of the optical fibre properties in exactly the right way to balance the dispersion of the pulse. Laboratory experiments confirmed that solitons could be used for ultra-high-speed communications where billions of solitons per second carry information down immensely long optical fibre links. For example, in 1999 the NTT laboratories in Japan demonstrated transmission of solitons at 10 Gigabits per second over a fibre-optic length of 180 million kilometres, and at 80 Gigabits per second over 10,000 kilometres. Recirculating solitons in tiny optical 'microresonators' have been used to generate a 'comb' of frequencies to incredibly high precision. Such frequency combs can be used to generate closely-spaced optical carriers with an immense combined data capacity - researchers have used 179 optical carrier frequencies to transmit 55 Terabits of data per second over a distance of 75 km - the equivalent of two million HD TV channels. Optical frequency comb generation has become an important scientific tool, one that enabled the measurement of the spectral line wavelengths of hydrogen to an unprecedented precision of 1.4 parts in 10^{14} - for which Theodore Hänsch and John Hall shared the 2005 Nobel Prize in Physics. John Scott Russell's acute observation of his 'heap of water' has had implications that have rippled far and wide indeed.

Back in the 19th century, Russell's immediate concern was in understanding how his wave of translation could be used to optimise the design of ships. He reported his original observation in a paper presented in 1835 to the Bristol meeting of the British Association for the Advancement of Science, and showed how the wave of translation could be used to reduce the water resistance to vessels moving fast in a restricted waterway. The interest was so great that Russell and Sir John Robinson, secretary to the Royal Society of Edinburgh, were appointed to carry on the investigations into the whole subject of waves, at the Association's expense. They reported back after two years, with three wide-ranging papers, including : '*On the Mechanism of Waves in Relation to the Improvement of Steam Navigation*', in which Russell described a new approach to optimising the design of ship hulls, which he called his 'wave-line' theory. In essence, this proposed that the profile of ship hulls should resemble the shape of the bow waves they created, with slim, concave bows shaped as sinusoidal curves to push water aside with minimum energy. This was a semi-empirical deduction, rather than a mathematically rigorous one, but it had a strong influence on subsequent hull design, as in the fast 'clipper ships' of the 1840s and beyond. Finding an entire hull shape to minimise water resistance was a problem that would be solved by William John Macquorn Rankine and William Froude many years later. Working at the Greenock shipyard of Thomson and Speirs, Russell introduced his wave-line designs to a series of vessels, including the *Skiff, Wave, Storm* and *Scott Russell,* followed by four fast Royal Mail ships, the *Teviot, Tay, Clyde* and *Tweed*. As well as their innovative, streamlined hulls, Russell introduced new structural designs for the iron ships, involving a system of longitudinal girders combined with numerous transverse bulkheads and a continuous iron deck; in effect, a box-girder construction conferring great strength and stiffness to the ships.

In 1836 Russell married Harriette Osborne, the daughter of Daniel Toler Osborne, an Irish baronet, and in the course of the next few years they had two sons and three daughters. His reputation as a foremost naval architect was established, and in 1844 Russell relocated his family to London, where he became engaged in a range of writing, editorial and engineering projects. He was invited by the Society of Arts to be its joint secretary, and helped to initiate its proposals for a Great Exhibition in what would become the marvellous Crystal Palace. In the aftermath of the Exhibition's great success, the dismantling of the Crystal Palace, and its reconstruction at Sydenham, in the south of London, the Prince Consort, Prince Albert, wrote that many difficulties had been encountered, and it was *"by dint of Mr. Scott Russell's tact, judgment, penetration, resource and courage, that obstacles vanished and intrigues were unmasked."*

Throughout his life Russell's main preoccupation remained firmly in naval architecture and shipbuilding, and in 1847, with partners, he acquired the Fairbairn shipyard at Millwall, and constructed the wave-line based yacht *Titania* for the English railway engineer Robert Stephenson, and *Adelaide* and *Victoria* for Isambard Kingdom Brunel and the Australian Royal Mail Steam Company. *Titania* enabled Stephenson to join the Royal Yacht Squadron, which had invited the New York Yacht Club's own wave-line based *America* to compete at Cowes for their Hundred Guinea Cup. In August 1851 *America* resoundingly defeated a flotilla of 14 British boats in the race that became the America's Cup. In the 'London Journal' a cartoon showed Queen Victoria asking which yacht came second, and being told *"Ah, your Majesty, there is no second."* A week later *America* raced *Titania* head-to-head in a battle of wave-line designs and won again. Russell acknowledged the victory graciously.

Meanwhile, ever ambitious, Brunel was hatching plans for his most audacious project yet. In 1852 he began sketching designs and calculations for an enormous ship capable of carrying its own fuel for an uninterrupted round trip from Britain to the Far East. In the finalised design the ship would measure 692 feet in length by 83 feet in the beam, displace 27,000 tons, and carry 4,000 passengers, 3,000 tons of cargo and over 10,000 tons of coal. To finance the project Brunel approached the Eastern Steam Navigation Company (ESNC) which had just competed unsuccessfully to win mail contracts to the Far East. Surprisingly, the beleaguered company accepted Brunel's bold vision and appointed him its chief engineer. To build the great ship, to be called the *Leviathan*, quickly became Brunel's major obsession and the ESNC's critical project. They invited tenders to build the ship, and received only one to build the whole ship, including engines, from Scott Russell's Millwall yard. Russell offered to build the hull for £275,200, with the paddle engines and boilers for £42,000 and the propeller screw engines and boilers for £60,000, sub-contracted to James Watt & Co. of Birmingham. This total was a great under-estimate compared to Brunel's private estimate of £500,000 to build the whole ship. The sole tender was accepted but the project soon ran into problems.

Construction started well and with high quality. The hull embodied the well-established Russell philosophy of a wave-line form, longitudinal iron stringers and strong bulkheads. But the sheer scale of the ship required Scott Russell's company to hire the neighbouring Napier yard to accommodate the work, and Brunel's frequent design changes increased the costs. By 1856, J. Scott Russell & Co. were insolvent and the ESNC took control of the yards to complete the ship, now becoming known as the *Great Eastern*. She was eventually launched, after many issues and a fatal accident related to her huge size and weight, in January 1858. The construction had cost £600,000 to that point, contriving the difficult launch had cost £120,000 and another £200,000 was required for the engines and fitting out. The ESNC was forced into liquidation with huge losses for its shareholders, including Brunel, and sold the incomplete ship for £165,000 to a newly formed 'Great Ship Company'. Scott Russell had managed to resurrect his own business and won the contract to complete the fitting out. The *Great Eastern* sailed for the first time in September 1859 and almost immediately, a stopcock left accidentally closed caused an explosion which killed five stokers. Brunel, already terminally ill, died several days later and never saw his and Russell's 'great babe' at sea.

The *Great Eastern* was never a commercial success apart from her use in laying sub-oceanic telecommunication cables. But she was a masterpiece of pioneering design, too far ahead of her time. It would be another 49 years before a larger ship was constructed: the Clyde-built Cunard liner *Lusitania*. The first failure of Scott Russell's shipyard initiated a series of financial problems for him. Involvement in a business to supply guns to the American Civil War also left him heavily indebted and much of his property was sold. The bankruptcy of his son's shipyard on the River Taff in 1869 caused him further losses.

Russell's later career was spent as a consulting engineer producing designs such as the Great Rotunda for the Vienna Exhibition of 1873, then the world's largest clear-span roof. He prepared some designs for the 1,000-foot span required for the London Tower Bridge, but while investigating potential ironworks for the project, he became ill and eventually died aged 74, in 1882.

In July 1995, an international group of scientists attended a conference on 'Nonlinear waves in physics and biology' at Heriot-Watt University, and gathered on the Union Canal near Edinburgh to witness a re-creation of Russell's famous first sighting of a solitary water wave. The occasion marked the naming of the new 'John Scott Russell aqueduct' which now carries the Union Canal over the Edinburgh City Bypass.

33 Light from oil: James 'Paraffin' Young (1811-1883)

From the Middle Ages to the middle of the 19th century, candles were the standard – sometimes the only - way of lighting homes and workplaces. Candlemakers, called 'chandlers', were members of skilled craft guilds, though their manually-produced products were far from perfect. Most early candles were made of tallow (animal fat), which burned smokily with unpleasant smells, and the superior alternative beeswax candles were expensive. Two innovations helped. In the 1820s, a French chemist, Michel Chevreul,

Illustration 38: James Young, chemist

discovered how to extract stearic acid from animal fat and make stearic wax, which burned durably and cleanly; and in 1834, the English inventor Joseph Morgan developed a machine for the continuous production of moulded candles. With mechanisation, candles became an affordable commodity for the masses. In his 1848 Royal Institution Christmas lectures for young people, Michael Faraday, fascinated by all aspects of science, discussed, with demonstrations *'The chemical history of a candle'*. He said *"Here are a couple of candles commonly called dips. They are made of lengths of cotton cut off, hung up by a loop, dipped into melted tallow, taken out again and cooled, then re-dipped until there is an accumulation of tallow round the cotton...The candle I have in my hand is a stearin candle, made of stearin from tallow in the way I have told you. Then here is a sperm candle, which comes from the purified oil of the spermaceti whale. Here also are yellow bees-wax and refined bees-wax, from which candles are made. Here, too, is that curious substance called paraffin, and some paraffin candles, made of paraffin obtained from the bogs of Ireland."* Paraffin wax, which burned cleanly and without odour, was widely introduced in the 1850s, after the Scottish chemist James Young found how to separate and refine paraffin from crude petroleum. Paraffin wax proved more economical to produce than any other candle fuel.

Young was born in Glasgow and educated himself at night-school in Anderson's University, where he attended the chemistry lectures of Thomas Graham. He became Graham's laboratory assistant, then began giving some of Professor Graham's lectures himself. At the 'Andersonian' he met many sons of industrialists and made life-long friendships with fellow students David Livingstone, a medic whose later African explorations he would support, and Lyon Playfair, who would become professor of chemistry at the Royal Manchester Institution and then the University of Edinburgh, and later the Liberal Member of Parliament for the universities of Edinburgh and St Andrews. When Graham left in 1837 to take the chair of chemistry at University College London (UCL), he took Young with him as an assistant, joined also for a while by Playfair. The next year Young, age 28, married his cousin Mary Young, which prompted, to Graham's disappointment, Young's move from academia to better-paid work in industry. He became manager, then technical director of the James Muspratt & Sons chemical manufacturers in Merseyside, owned by the father of another Andersonian student, Sheridan Muspratt, who was also one of the assistants Graham recruited from Glasgow to London. Young's first patent was for the production of ammonia, building on development work he had done at UCL, using carbonic acid. He had a spiky relationship with James Muspratt and in 1844 moved onwards and upwards to a lucrative job as a semi-independent consultant at

Tennants, Clow & Co in Lancashire, producers of acids, alkalis, bleaching agents and dyes, and a subsidiary of Tennants of Glasgow. For them, he developed several patented process improvements for the production of indigo dyes, copper sulphate and stannate of soda and potash, (receiving a third of profits as his royalty) and he surveyed Northern England for potential sites of new chemical works, identifying in particular, Middlesborough. Around this time, Young's political views, which were radically Liberal, caused him to join the Manchester Anti-Corn Law League and he helped to found the *Manchester Examiner,* intended to be a more radical alternative newspaper to the 'Gladstonian' and 'Peelite' *Manchester Guardian.*

Young was reacquainted with his old friend Lyon Playfair who was professor at Manchester's Royal Institution from 1843 to 1845. Playfair had begun to improve the "chemical life" of the city, first by holding informal meetings at his home at Whalley, and then by forming, with Young, a chemistry section within the Manchester Literary and Philosophical Society. Young and John Thom, yet another Andersonian student of Thomas Graham, were asked by the Society to investigate causes and possible solutions to the Irish Potato Blight *"for the purpose of ascertaining how far chemical agents could be used in averting the disease."* Thom and Young showed that sulphur vapour was an effective protection against the blight and that immersing diseased potatoes in weak sulphuric acid solution halted the progress of disease. The information was passed on to Prime Minister Peel, but he was intent on a political solution - in the form of the repeal of the Corn Laws, meaning the abolition of tariffs on the import of grain. The Corn Laws were repealed, but the political action was ultimately ineffective in preventing the great Irish Famine.

In December 1947, Playfair wrote from London to Young, to tell him of a wellspring of oil which was yielding 300 gallons per day at a colliery in Derbyshire owned by his brother-in-law. Playfair wrote that the 'naptha' was rather like treacle and after one distillation *"...it gives a clear, colourless liquid of brilliant illuminating power."* No doubt aware of the work of Charles Macintosh, he added that *"it dissolves caoutchouc easily."* Young visited the site in the same month and began the experimental distillation of the crude petroleum. On New Year's Day, 1848, he found that some pitch he had purified *"had a crystalline appearance on top fluid with small grains as if it had little crystals in it."* These were tiny flakes of the wax named 'paraffine' by the German chemist Carl Reichenbach. Young reported to Tennants the potential to refine the oil. They considered the likely scale of production to be too small, but agreed that Young could establish his own refinery under concession from the landowner, while remaining employed as their technical consultant. Young set up a small refining business and successfully distilled a thin light oil suitable for lamps, and also produced a thicker oil good for lubricating machines. The main customers were textile manufacturers who appreciated the cost-effective lubricant, but as perhaps anticipated by Tennants, the supply of oil from the mine soon began to dwindle. Searching for a solution, Young noticed some oil dripping from the sandstone in the roof of the mine, and concluded (mistakenly) that it has been formed by the action of heat on the coal seam. He began to experiment in distilling coal, from many sources, at various temperatures. The Andersonian graduate network was useful again, and he was sent a barrel of 'cannel coal' by Hugh Bartholomew, manager of the City and Suburban Gas Company in Glasgow. The cannel (candle) coal was mined from the Torbanehill Estate coal seams near Bathgate, and was also known as torbanite or Boghead coal. It was a type of oily coal that burned brightly, with a long, luminous candle-like flame, and was akin to oil shale. After months of experiments Young wrote at last that *"out of a cannel that came to be mixed with soda ash for making the alkali, I got a quantity of a liquid that contained paraffine."* Young had identified the general type of coal most likely to yield crude paraffin oil in great amounts, and was fairly certain of the best temperature for the distillation. After a thorough patent search and literature review, he submitted and was awarded, a patent claim in October 1850 for *'Treating Bituminous Coals to Obtain Paraffine and Oil containing Paraffine therefrom'*. During the winter of

1850-51 Young specified and commissioned the world's first commercial oil plant, and sited it two miles from the Boghead coal pithead in Bathgate. After a year, oil production began. Young's business partners were Edward W. Binney, a Manchester lawyer and amateur geologist, expert in patent law; and Edward Meldrum, Young's technical assistant who had managed the pilot production in Derbyshire.

The first products were naptha, used as a solvent; 'Blue Oil', for lubrication; and a refined lubricating oil that Young called 'Finished Liquor'. These names and others including 'Black Liquor' and 'Green Liquor' were used to disguise the nature of the products and gave nothing away about their method of production, but they became standard technical terms used throughout the oil industry. The secrecy was compromised by the need to reveal technical details in court, after litigious attacks on the validity of the patent by the owners of the Torbanehill Estate, who claimed that torbanite was not coal, and therefore not included in their tenant's lease agreement. Their claim was rejected, but the large profit margins being made by Young & Co. were revealed in court and further legal actions were required to defend the patent against other eager competitors. By 1855, Young's most profitable product was paraffin for lamps, despite the increasing take-up of coal gas lighting in British city streets and wealthier homes (as pioneered by William Murdoch from Ayrshire who had gas-lit his own house in Cornwall as early as 1794 while working for Boulton and Watt). Paraffin wax was in production the next year and was selling in quantity for candles by 1859. American kerosene producers began using Young's process and he spent three months in the United States to ensure his licence royalties of 2 cents per gallon.

In a few years the supply of Boghead cannel coal was nearing exhaustion and its price had more than doubled from 13s-6d to 30s per ton. Young started exploring the possibility of replacing it with oil shale, a black, laminated rock that can be cut with a sharp knife, which can produce oil when heated to between 350 and 500°C. He personally acquired land on several promising sites and after several disappointing surveys like one in the west of Glasgow, *("Looked at a bed of shale about 6 feet thick in a quarry at the Botanic Garden end of Byres Road - it is a common shale won't burn ...did not distill it as it is hopeless..")* Young identified a usable shale deposit not far from Bathgate at Addiewell, West Calder. When his coal-distillation patent expired in 1864, he bought out his partners in the Bathgate works (for the sum of £32,000) and established an oil shale works at Addiewell, setting up with £600,000 of investors' capital, as Young's Paraffin Light and Mineral Oil Company. His friend David Livingstone laid the foundation stone of the works, and a brickworks was established to build the 420 houses needed for the oil-plant workers. Young guaranteed to his new investors (who included Lyon Playfair and Hugh Bartholomew) that the oil-shale in the land he held either as landlord or lessee would produce a total of 30 million gallons of crude oil, in return for fees and fixed rents on his land. In the first few years of operation, trading conditions were difficult. Crude oil was being imported from the United States and Burma, and refined by George Shand & Company at the Forth Bank Chemical Works near Stirling. Oil prices fell by a half and Young's board of directors argued about the value of the assets he had transferred to the Company. During a period of 'Scottish oil mania', many shale-oil companies had been formed and now several were going bust. The completion of the Addiewell plant was delayed, and Young's decisive and action-oriented temperament was proving to be unsuited to management by executive committee. Friction arose between factions and especially between Young and John Moffat, a major shareholder. Nevertheless in 1866 the Company's works produced 3.2 million gallons of crude oil. By 1868 trade was much healthier: the Addiewell works were completed, and produced 6 million gallons of crude, with demand outstripping supply. Soon the Bathgate and Addiewell operations were the largest employers in the area.

Young had made his fortune, and when he fell ill in late 1869, he took a Mediterranean cruise and decided that he had had enough of the board of directors. He resigned, while maintaining his financial interests in the Company. In any case, he had other interests. Young had become a Trustee directly involved in the administration of his *alma mater*, Anderson's University in Glasgow, and was a passionate advocate of adult technical education. He had paid the wages of a skilled mechanic to assist the Anderson's professor of Natural Philosophy for many years, and in 1868 was unanimously elected President of Anderson's University, a position of great authority in the management structure designed by John Anderson. Through personal grants and solicited fund-raising he rid the Andersonian institution of debt, and endowed the Young Chair of Technical Chemistry together with its buildings and laboratories. At Young's instigation a course of evening lectures on marine engineering began and was a great success. Under his stewardship the value of the institution's assets rose from £5,000 in 1868 to £40,594 in 1872, and over his tenure he contributed more than £27,500 to Anderson's University. The next problem to be addressed was the legal status of the institution. Its Trustees were not incorporated, so their funds, and those of the University, were inadequately protected. And with no Royal Charter, its status as a university was challenged, especially by the Senate of the University of Glasgow, who had no more love now for the late John Anderson and his rival institution, than they had for John Anderson when he was alive. James Young lobbied the 'Glasgow Commission for the Promotion of Technical Education' saying *"...None of our universities or Institutions have hitherto provided the kind of instruction required - in a great measure in consequence of the difficulty of finding teachers who possess a practical knowledge of...the application of science..."* The need was a national one. In Germany and Switzerland, technical universities in the form of *Technische Hochschulen* had been created. In France, the *École Polytechnique* and *Grandes écoles d'ingénieurs* were regarded enviously by British engineers and scientists. Yet a generation earlier, in 1825, Baron Dupin, in advising the French government on *'The Commercial Power of Great Britain'*, had pointed to the Andersonian Institution as one of those *"...which conduce to the progress of industry and of the knowledge that directs it"* and in admiring Anderson's management structure noted *"Be it remarked, that the government has no concern or influence directly or indirectly with this management !!!* 'Since then, the international competition had raced ahead. Among others, Lyon Playfair and John Scott Russell also saw the gap, and joined the campaign to establish technical universities across Britain. In response, the government appointed a Royal Commission to investigate.

To gain the desired Royal Charter, the managers of Anderson's institution decided to focus on the 'Royal Commission on Scientific Instruction in Theoretical and Practical Science to the Industrial Classes', optimistic that the Liberal government of Gladstone was well-disposed to educational reform. To the Commission, Young proposed that government grants should be made to technical universities across Britain, saying *"I cannot see a better application of money than by teaching science..."* and observing that the lack of suitable endowments had greatly hindered the work of Anderson's University: *"We have been a training school for professors. London and the University of Glasgow have taken some of our best..."* Opposition to the grant of a Royal Charter came predictably from the Glasgow University senate, and the Royal Commission did not recommend the grant. Young then applied instead for a Private Bill of Incorporation, and by 1876 the parliamentary solicitors had prepared a Bill to be heard by the House of Lords. Again opposition came from the Senatus of Glasgow University, who petitioned against the Bill. They objected to a name containing the words 'University' or 'College', preferring the title to be 'Academy of Science' or 'Institute of Science'; wanted the professors to be called 'lecturers'; and opposed the right of the institution to award degrees. This last restriction would actually cause little problem for most of the Andersonian's students, (except for its medical students, who needed their qualifications to be nationally recognised). After all, the *Technische Hochschulen* did not grant degrees either. A compromise was reached. The 'Anderson's College (Glasgow) Act' received Royal Assent in 1877. It changed the name of the institution, incorporated the Trustees, and modified their role together with that of the managers, so they were no longer disqualified by association with Glasgow University or the city council. Young's Chair of Technical Chemistry stayed within the College, and the managers would have full powers to appoint and dismiss professors. The medical faculty was separated and incorporated to become Anderson's College Medical School. Having completed the task of reorganising and refinancing the Andersonian, James Young retired as President of the renamed Anderson's College. It would become the West of Scotland Technical College, then the Royal Technical College, then the Royal College of Science and Technology, and finally the University of Strathclyde.

With two other trustees, Young had managed the affairs of David Livingstone while the explorer was in Africa. They raised funds and donated money to support the expeditions, and Young acted as guardian and mentor to the unsettled Livingstone children. Dr. Livingstone's correspondence was as regular as it could have been. In 1858, sailing to Africa aboard the steamer *Pearl* with the parts of his prefabricated exploration boat on the deck, he wrote to Young whimsically, *"Here we are off Cape Corrientes (Whaur's that I wonner?) and hope to be off the Luabo four days hence. We have been most remarkably favoured in the weather, and it is well, for had our ship been in a gale with all this weight on deck, it would have been perilous... I met a Dr. King in Simonts Bay, of the 'Cambrian' frigate, one of our classmates in the Andersonian ..."*

In his later years, James Young lived at his estate near Wemyss Bay in Ayrshire, while farming his three other estates spread across central Scotland. On his impressive steam yacht, the 214-ton teak-and-iron vessel *Nyanza*, named after Livingstone's explorations around Lake Victoria, he sailed as far as the Mediterranean, Denmark, Iceland, and the Norwegian fjords. Young continued experimenting, inventing and producing patents, and remained prominent in the British Association for the Advancement of Science and in the Chemical Society. He gifted memorial statues of Livingstone and Thomas Graham to the city of Glasgow. In 1883, Young died at home aged 71, survived by four daughters and three sons, and left an estate of more than £165,000. Young's Paraffin Light and Mineral Oil Company continued in production, and by the 1900s it was employing 4,000 people and extracting around 2 million tons of shale every year. Young and his companies were instrumental in founding the modern petrochemical industry. In 1919, the Paraffin Light and Mineral Oil Company was merged to form part of Scottish Oils Ltd, a subsidiary of the British Petroleum Company (then the Anglo-Persian Oil Company) - and thus BP became firmly established in Scotland.

34 Childbirth and anaesthesia: James Young Simpson (1811-1870)

Illustration 39: James Young Simpson

"Abundant evidence has of late been adduced, and is daily accumulating, in proof of the inhalation of sulphuric ether being capable, in the generality of individuals, of producing a more or less perfect degree of insensibility to the pains of the most severe surgical operation. But whilst this agent has been used extensively, and by numerous hands, in the practice of surgery, I am not aware that any one has hitherto ventured to test its applicability to the practice of midwifery...

"Within the last month I have had opportunities of using the inhalation of ether in the operation of turning, in cases of the employment of the long and short forceps, as well as in several instances in which the labour was of a natural type, and consequently required no form of artificial aid."

So wrote James Young Simpson to Edinburgh's *Monthly Journal of Medical Science* in February, 1847. He went on to describe several groundbreaking cases of using ether in difficult, forceps-assisted deliveries, and crucially concluded *"In all of them, the uterine contractions continued as regular in their occurrence and duration after the state of etherization had been induced ..."*

Simpson was 36 years old, in his eighth year as Professor of Medicine and Midwifery at the University of Edinburgh, and had long been concerned with the problem of pain in surgery and childbirth. He was born in Bathgate, the youngest of eight children of the local baker, David Simpson and his wife Mary. James was enrolled at Edinburgh University aged 14, following his friend and schoolfellow John Reid, who was two years older and later became professor of anatomy & medicine at the University of St Andrews. After the usual general classical studies, Simpson became a medical student, influenced by Reid, who zealously attended Robert Knox's famous anatomy lectures twice a day. Simpson's mother had died when he was aged nine, and his father David died while James was still a medical student, in 1830, at a critical point, about to take his surgeon's degree. Despite the grief and disruption of his father's death, Simpson passed with credit and became a member of the Edinburgh Royal College of Surgeons, qualified to practice medicine, when he was only 18 years old.

With practical, emotional and financial support from his elder brother Alexander (known as Sandy), Simpson managed to return to Edinburgh for further studies, and received the degree of Doctor of Medicine in 1832 for his thesis, defended verbally entirely in Latin, *De causa mortis in quibusdam inflammationibus proxima*', on the causes of death from certain inflammations. The professor of pathology, John Thomson was so impressed that he appointed Simpson as his assistant. It was Thomson who encouraged Simpson to attend the lectures of James Hamilton, Edinburgh's professor of midwifery, a profession whose 'man-midwives' or 'accoucheurs' were still derided and sneered at by many other medical practitioners. James Hamilton's father and predecessor, Alexander Hamilton, had been one of the first, after Alexander Gordon of Aberdeen, to recognise that puerperal (childbed) fever was caused by infection. Yet it would be many years before the germ theory of disease was widely accepted in operating theatres and delivery rooms. Hamilton's midwifery lectures led to Simpson's lifelong study of obstetrics, beginning with his inaugural paper on the diseases of the placenta, read to the Royal Medical Society of Edinburgh, of which Simpson was elected President in November 1835. Talented, ambitious and youthfully energetic, Simpson's medical career advanced rapidly, and many of Edinburgh's medical professors were nearing the end of theirs. In 1839, John Thomson (pathology) was 74, James Hamilton (midwifery) was 72, Monro *tertius* (anatomy) was 66 and Charles Bell (surgery, and who described the eponymous Palsy) was 65. That year, Hamilton resigned and soon died, and Simpson announced his candidacy for the position. On Boxing Day, he married his fiancée and second cousin, Miss Jessie Grindlay from Liverpool, which removed bachelorhood as one of the two obstacles to his election as professor of midwifery. The other one was his relative youth. After a vigorous campaign against the opposition of several Edinburgh professors, Young Simpson, living up to his name and aged only 28, was narrowly elected by the Town Councillors to the professorship over his rival, Dr. Evory Kennedy of Dublin, in January 1840. Simpson had succeeded to the chair of midwifery which, to the credit of the University and Town Council of Edinburgh, had been the first to be established in Great Britain and probably the world, in 1726. He wrote *"I was this day elected Professor. My opponent had sixteen and I had seventeen votes. All the political influence of both the leading Whigs and Tories here was employed against me; but never mind, I have got the chair in despite of them, Professors and all. Jessie's honeymoon and mine is to commence to-morrow."*

From Simpson's earliest days he was concerned to minimise patients' pain, and like many medical students he had been so shocked by observing surgical operations that he had almost abandoned ideas of becoming a doctor. Operating theatres were like butcher shops. Cleanliness was not considered important, and without anaesthesia the patients were just constrained by straps or several pairs of powerful arms while the surgeon did his work quickly above their screams, until they sometimes mercifully fainted. Occasionally hypnosis ('mesmerism') of the patient was tried, but with variable results. Experiments with diethyl ('sulphuric') ether as a surgical anaesthetic began in the United States, first by surgeon Crawford Long and then by dentist William T.G. Morton, whose demonstration of its effect at the Massachusetts General Hospital in the autumn of 1846 impressed surgeons and students alike.

News of the successful experiments spread quickly and the Edinburgh surgeon, Robert Liston, whose lightning-quick but savage operations had shocked the student Simpson, performed the first amputation under ether anaesthetic at the University College Hospital, London in December 1846. Liston enthused *"This Yankee dodge beats mesmerism hollow!"* Within a month Simpson used ether anaesthesia for the first time in obstetric procedures, and showed that its effects could be maintained for many hours. For the next year he used ether routinely in his obstetric practice and promoted its general adoption.

Simpson recognised too that the ether had its problems. It required large quantities, smelled unpleasant, caused occasional irritation to the bronchial passages, raised the patients' heart-rates, and sometimes induced anxiety or mania. Moreover, ether gas is dangerously flammable and potentially explosive. Simpson searched for an alternative, and a Liverpool chemist called David Waldie, originally from Linlithgowshire, suggested that he should try chloroform (trichloromethane). Simpson took the suggestion literally. He and his two assistants, George Keith and James Duncan, were in the strange, hazardous, late-evening habit of gathering in Simpson's dining room to inhale various chemicals to test if they had any anaesthetic effects. When they inhaled chloroform, they became very happy and talkative, and considered their conversation to be of exceptional intelligence - briefly. According to Simpson's neighbour and surgeon colleague, Professor James Miller, *"On awakening Dr. Simpson's first perception was mental - 'this is far stronger and better than ether' ...His second was to note that he was prostrate on the floor... Dr. Duncan beneath a chair...quite unconscious, and snoring...Dr. Keith's feet and legs making valorous attempts to overturn the supper table. By and by...each expressed himself delighted with this new agent."* One of the ladies present, Miss Petrie, a niece of Mrs. Simpson, gamely volunteered to take her turn and fell asleep crying *"I'm an angel! Oh, I'm an angel!"* The doctors were unanimous - something had been found that surpassed ether.

The next morning Simpson ordered a manufacturing chemist to make enough chloroform to supply his large midwifery practice and just a few days later, on November 10th 1847, he reported the initial results in a paper read to the Edinburgh Medico-Chirurgical Society (quickly published as a pamphlet and in *The Lancet*). He lauded chloroform's advantages over ether: *"1. A greatly less quantity of Chloroform than of Ether is requisite...2. Its action is much more rapid and complete...3. Most of those .. have strongly declared the inhalation of Chloroform to be far more agreeable..."* concluding that *"I have never had the pleasure of watching over a series of better and more rapid recoveries; nor once witnessed any disagreeable results follow to either mother or child; whilst I have now seen an immense amount of maternal pain and agony saved by its employment...I feel assured, that the use of Chloroform will soon entirely supersede the use of Ether."* By this time Professor Miller had successfully performed the first trials in surgery using chloroform anaesthesia.

Simpson was proved to be slightly over-optimistic in his enthusiasm, and he encountered particular opposition when he practiced and promoted the use of anaesthetics to manage the pain of natural childbirth. Many medical and clerical commentators took the view that the pains of 'parturition' were essential, character-building, or even divinely sanctioned. To be fair, the medical objections were mainly based on the fear that anaesthetics would increase surgical mortality (already of course very high). To test these objections Simpson gathered statistics from British, Irish and Parisian hospitals, on the outcomes of one of the most dreaded operations - amputation of the thigh. The returns from more than 1,600 cases in 50 hospitals showed that without anaesthesia the average frequency of death was 45%, and using anaesthetic, 25%. With the religious objectors, Simpson had little patience. Although he was an adherent of the Free Church of Scotland, he had refused to sign its commitment to the Westminster Confession of Faith, citing its literal interpretation of the book of Genesis. A leading exponent in the Dublin School of Midwifery objected that the feeling was very strong against the use of anaesthetic in normal childbirth, merely to avert the ordinary amount of pain, which the Almighty had wisely seen fit to allot to natural labour. Simpson replied sarcastically, *"I do not believe that anyone in Dublin has yet used a carriage in locomotion; the feeling is very strong against its use in ordinary progression, merely to avert the ordinary amount of fatigue which the Almighty has seen fit - and most wisely no doubt - to allot to natural walking...*" Most objections gradually faded away and were eventually overcome. The first natural birth delivered by Simpson, using chloroform, was the daughter of Jane Carstairs and christened Wilhemina, although Simpson nicknamed the baby 'Anaesthesia'. In 1847 Simpson was appointed a physician in Scotland to Queen Victoria, and he considered his triumph in the anaesthetic arguments complete when in 1853 he received a letter from Sir James Clark, physician in ordinary to Her Majesty, reporting that she had given birth to Prince Leopold, while under the influence of chloroform administered by Dr. John Snow.

Although it was the introduction of anaesthetics that made Simpson internationally famous, he made many more major contributions to obstetrics.

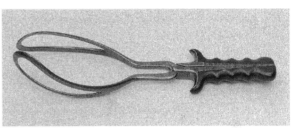

Illustration 40: An example of 'Young's Forceps'

He refined the design of forceps by giving them an elongated 'cephalic curve', specially shaped to fit the baby's head when it is temporarily elongated within the birth canal, and which are still widely used as 'Simpsons Forceps'. As an alternative to forceps he invented an 'Air Tractor' which used a suction cup attached to the baby's head to assist delivery. The method did not catch on but was resurrected a century later with the invention of the 'ventouse'. Throughout his career he was noted for his thoroughness, sympathy and gentleness in dealing with his patients, and his children. He was a champion of breastfeeding and the importance of the mother-child 'bonding' process. He often gave his services for free. But in promoting his scientific approach and in exposing quackery, he was relentless.

Homeopathic practices had been growing in Edinburgh, and even the professor of pathology, Henderson, (who owed his appointment largely to Simpson's support) became enamoured. Simpson was appalled and enraged. Joined by fellow professors, James Syme and Robert Christison, he championed the cause of evidence-based medicine, and Henderson was forced to resign 'on loss of health'. On mesmerism, Simpson was more ambivalent. Accused of being a supporter in a London journal, he repudiated the suggestion and proposed to settle the matter by experiment. He offered five sealed boxes each containing a line of Shakespeare he had written, and £500 to any clairvoyant who could read the lines. The offer was not accepted. If the feat had been achieved, of course, that would have been mind-reading, not hypnotism.

In 1866 Simpson was knighted, the first knighthood conferred on a doctor in Scotland. He had twice before refused to take a title, but accepted the honour this time and enjoyed the crowds who gathered in Queen Street to congratulate him. Two years later, the Principal of Edinburgh University, Sir David Brewster, died and Simpson was urged to stand for election. Not wanting to resign his professorship, he declined and recommended Christison, who in turn said that Sir Alexander Grant, a historian and businessman, would be a better choice. Grant was duly elected.

Throughout his career Simpson was concerned by the problems of hospital infection and surgical sepsis. He had long supported Hamilton's view that puerperal and surgical fever were identical and that it was proved by evidence that sufferers of the childbirth disease had been *"locally inoculated by a materies morbi capable of exciting puerperal fever; that this materies morbi is liable to be inoculated into the dilated and abraded...maternal passages, during delivery, by the fingers of the attendant"*. Simpson quoted the statistics gathered in Vienna by Ignaz Semmelweis: of 21,120 women delivered from 1840 to 1846, about 1 in 10 died, chiefly from childbed fever, having been examined multiple times by medical students. After 1847, with the prevention of patient touching by medical students and the introduction of chlorine handwash, the mortality reduced to 1 in 74.

Simpson was scathing about attitudes and practices on the continent of Europe: *"...was not the want of a due knowledge of the communicability of puerperal fever the cause immediately leading to these numerous maternal deaths? ....sacrificed merely to medical prejudice, in the form of a total disbelief...in the contagious communicability of puerperal fever? Is not the very high mortality seen in most continental lying-in hospitals, as compared with those of Great Britain, a result principally of inattention to the doctrine of the communicability of puerperal fever? And, lastly, if puerperal fever may be occasionally communicated to puerperal patients, may not surgical fever be...in the same way?"* Simpson had no doubt that it would take many long years to fully convince surgeons of the communicability of surgical fever. Simpson rammed home the lesson with more statistical surveys, and in two papers published in 1869 coined the term *'hospitalism'* to describe the dangers of hospital confinement, compared to operations performed in rural homes. Assuming the surgical skills of the practitioners in town hospitals and rural homes to be the same, he analysed the influence of the surrounding environment on the patient, for more than 4,000 limb amputations, divided almost equally between rural practices and city hospitals. The results were startling. *"After the 2,098 limb-amputations in the country, 226 of the patients died. After the 2,089 limb-amputations in eleven large metropolitan hospitals, 855 of the patients died. The mortality...in the country is thus 1 in 9.2, and ...in large and metropolitan hospitals 1 in 2.4... nearly FOUR TIMES GREATER."* As a consequence, Simpson advocated changes in the design of hospital layout and management towards small isolated units - influencing many later practitioners including Florence Nightingale. Simpson, incidentally, supported medical education for women. Strangely though, he was sceptical of the work of Joseph Lister at Glasgow, who in 1867 showed the effectiveness of antisepsis in surgery, believing that the answer to surgical sepsis lay in improved hospital design and practice, not carbolic dressings.

In 1870 Simpson died, aged 59, at home in Edinburgh, having suffered from chest pains. Sadly, five of his nine children had died before him. His family declined the offer of a grave in Westminster

Abbey and he was buried in his home city's Warriston Cemetery. A bust was placed in the Abbey to remember that for his *"genius and benevolence the world owes the blessings derived from the use of chloroform for the relief of suffering"*. Chloroform and diethyl ether continued in widespread anaesthetic usage for many decades until supplanted by better alternatives, such as nitrous oxide and fluranes, though ether is still used in some countries. The occasional side-effects of chloroform - cardiac fibrillation, liver toxicity, and possible carcinogenesis - became apparent in the 20th century, and its use declined from the 1940s.

35 Antiseptic surgery: Joseph Lister (1827-1912)

The magnificent building of the Glasgow Royal Infirmary, designed by architect Robert Adam, was opened in 1794 on a site near Glasgow Cathedral. With an initial 100 beds spread over five floors, two physicians and four surgeons, and a circular operating room with a glazed dome ceiling, it was a state-of-the-art institution, designed with care and experience. Adam's father, William, had designed the Royal Infirmary of Edinburgh, which had opened its doors almost fifty years before. In the 19th century, Glasgow's thriving industries meant that patient demand for the free medical care that the 'Royal' provided soon outstripped the capabilities of its original Infirmary building. In 1861 a new surgical wing was constructed with an additional 144 beds, bringing the Infirmary's total to 572. Yet problems remained. *'Operation successful, patient died'* was a common entry in medical records at Glasgow and elsewhere. James Young Simpson, the pioneer of anaesthetics, had pointed to the syndrome of 'hospitalism', writing *"A man laid on the operating table in one of our surgical hospitals is exposed to more chances of death than...on the field of Waterloo."* For patients, surviving surgery was only half the battle. And the brand-new surgical wing at Glasgow's Infirmary did nothing to reduce the death rate. Some of the superstitious wondered if

Illustration 41: Glasgow Royal Infirmary, c.1812

building the new wing so near to the old cathedral graveyard had angered dormant spirits.

The Regius Professor of Surgery at Glasgow University was a young man named Joseph Lister, aged 33 and a newly-elected fellow of the Royal Society, for his studies of inflammation and the mechanism of blood-clotting. He was practical as well as academic. As a clever but nervous boy he had been tutored at home in Essex by his prosperous wine merchant father, who was a self-educated polymath skilled in making and using microscope lenses. As a Quaker, Joseph was ineligible to study at Cambridge or Oxford, and enrolled instead at University College London. As a student he had been present at the first operation under ether conducted in Britain, at University College Hospital by Robert Liston, and he studied medicine at Edinburgh under the tutelage of Liston's protégé, professor James Syme, an inspiring teacher and 'the first of British surgeons'. Joseph had married Syme's daughter Agnes in 1856. When the Glasgow Royal needed a leading surgeon at the new facility, Lister was eminently qualified, available and near-at-hand - and he applied for a position at the Infirmary. His initial application, amazingly, was rebuffed. The chairman of the governing Board, one David Smith, said *"But our institution is a curative one. It is not an educational one."* Lister persisted, and after a year he was appointed to be surgeon in charge of the Male Accident Ward.

Many of Lister's male patients suffered from industrial injuries that often required limb amputations. When he analysed the statistics he was shocked to realise that between 1861 and 1865, nearly half of amputations resulted in death from postoperative sepsis - the main cause of hospitalism. No one seemed too concerned - this was normal and to be expected. But was this really acceptable, in a shiny new hospital unit, free from the 'bad-air' that was supposed to be the problem in older hospitals? Lister thought not, and began to research and investigate. He came across an 1864 paper in *The Lancet* describing the successful sanitisation of sewage by carbolic acid, in the town of Carlisle. As he was so concerned about the incidence of sepsis and gangrene in amputations, open wounds and fractures, he seized on the idea and wrote later in the same journal *"My attention having for several years been much directed to the subject of suppuration...I saw that such a powerful antiseptic was peculiarly adapted for experiments... and while I was engaged in the investigation the applicability of carbolic acid for the treatment of compound fractures naturally occurred to me."* Lister was also persuaded by the papers of the French physicist and chemist Louis Pasteur, *'Recherches sur le putréfaction'*, which were pointed out to him by his University colleague, the professor of chemistry, Thomas Anderson, as they were walking home together one chilly Glasgow night. From his studies of diseases of yeast in wine and beer, Pasteur concluded that specific microorganisms were responsible, and that the same mechanism could cause disease in animals and humans - the 'germ theory of disease'. Pasteur famously once said *"dans les champs de l'observation, le hasard ne favorise que les esprits préparés"* - in observing, chance favours only prepared minds - and Lister's mind was already fully prepared.

His first experiments with carbolic acid were in March 1865. Lister treated two patients with open fractures, one of the wrist, the other of the leg. Infection occurred in each case and amputations were required. He blamed *"improper management"* of these cases and over the spring and summer, he mulled over what needed to be done differently. In August, an 11-year-old boy was admitted with an open fracture of his left tibia, caused by the wheel of a cart running over the middle of his shin. After setting the fracture, the wound was treated with a dressing of lint dipped in carbolic acid, and after a few days washed with dilute carbolic. Success! There was no infection or suppuration of the wound. In a second case the next month, a 32-year-old labourer with an open fracture of the tibia, caused by a kick from a horse, was also treated successfully with the same technique - initially. On returning from a trip away from Glasgow lasting several weeks, Lister was *"deeply mortified to learn that hospital gangrene attacked the sore soon after I went away, and made such havoc that amputation became necessary."* The problem seemed to be that the carbolic acid was evaporating and leaching away from the conventional

Illustration 42: Joseph Lister c.1855

dressings covering the wound, and caused some blistering of the surface of the surrounding skin, giving another chance for infections to invade. Lister thought of a metal covering, made of thin sheet lead or tin, to contain the carbolic in the wound for longer. An initial test with a small wound and a lead covering about the size of a coin was successful. The chance to try the idea on a serious wound came eight months later in the form of a 21-year-old iron foundry moulder, injured by a falling iron box containing moulding sand, resulting in a horrible compound fracture of both the tibia and fibula of his left leg. After setting, Lister applied carbolic-soaked lint covered by a piece of sheet tin about 4 inches square. A larger knee-to-ankle sheet of tin was used the next day to replace splints, and a twice-daily regimen was started of hot compress washes and the application of more dilute carbolic, for three weeks. Six weeks after the accident, the bones were united, and the wound healed. Lister and his team went on to treat other even more serious open fractures with the method of carbolic dressings contained in tin shields, with success in cases where amputation of the limb would normally have been unavoidable. In a significant comment while debating whether carbolic mixed with olive oil was preferable to undiluted acid, Lister wrote *"Considering...that the patient's life may depend upon the entire destruction of the septic germs that lie in the wound, I am inclined to think it wiser to avail ourselves of the full energy of the pure acid..."* (my emphasis). Lister had completely accepted the germ theory of disease, even if his concept of 'germs' was not precisely that of later germ theory, since he believed that air was the sole source of germ infection. But realising the dangers of patient contamination in the operating theatre, he went further and instructed his surgeons to wash their hands before and after operations in 5% carbolic solution, to do the same with improved all-steel instruments, and to wear clean gloves, while during operations their assistants sprayed the carbolic solution around the room.

Lister's methods reduced the incidence of death from postoperative infection to the extent that his wards achieved nine months without losing a single patient to sepsis. His publications in *The Lancet* prompted an editorial in the issue of August 1867. *"If Professor Lister's conclusions with regard to the power of carbolic acid in compound fractures should be confirmed...it will be difficult to overrate the importance of what we may really call his discovery."* But like many pioneering discoveries, Lister's innovations were widely challenged. In 1870 Lister reviewed progress in the context of his experience at the Glasgow Royal, in another *Lancet* paper, *'On the effects of the antiseptic system of treatment upon the salubrity of a surgical hospital'*. Among the discoveries revealed by his record-keeping in the early days of the Royal's new surgical building, he had found that the highest rate of his patient mortality was on the ground floor, followed by the first floor, *"yet none of my wards ever assumed the frightful condition which sometimes showed itself in other parts of the building, making it necessary to shut them up entirely for a time."* Suspecting the cause of the problem to be a 'foul drain', an excavation was ordered, that *"disclosed a state of things which seemed to explain sufficiently the unhealthiness that had so long remained a mystery. A few inches below the surface of the ground, on a level with the floors of the two lowest male accident wards...was found the uppermost tier of a multitude of coffins, which had been placed there at the time of the cholera epidemic of 1849...the wonder now was, not that these wards ...had been unhealthy, but that they had not been absolutely pestilential."* The mass of bodies was too great to contemplate their removal. The corpses were 'freely treated' with carbolic acid and quicklime, and walls were rearranged to avoid the ingress of 'bad air' from the burial site. Lister pointed out the reduction in death rate (of around 65%) that his antiseptic system had achieved even in such an imperfect hospital site. Further evidence, if it was needed, emerged from the Franco-Prussian War of 1870, where the treatment of shrapnel wounds with carbolic acid by German doctors greatly reduced mortality from sepsis and gangrene.

Lister left the Infirmary and the University of Glasgow, at the comparatively young age of 42, to succeed his father-in-law James Syme as professor of clinical surgery at Edinburgh. In his inaugural lecture he said of the use of antiseptics, *"Surgery becomes something totally different from what it used to be; injuries and diseases formerly regarded as most formidable, or even hopeless, advance quietly and surely to recovery. Of this system the germ theory of putrefaction is the pole-star which will guide you through what otherwise would be a navigation of hopeless difficulty."* Lister became almost instantly popular with his medical students and he continued practical research to improve his surgical dressings and find even better antiseptic chemicals. He also used the power of microscopes to demonstrate the reality of microorganisms growing in fluids exposed to dust and air. Even then, one of his critical Edinburgh colleagues, professor of medicine Hughes Bennett, (supposedly an expert microscopist and seemingly also an expert in Egyptology) failed to replicate Lister's experiments, and ridiculed the germ theory. *"Where are the germs, then? Show them to us and we will believe....the dust has been ransacked for these organic germs...not only in frequented places but in desert places...in the ancient palace of Karnak, on the banks of the Nile; in the tomb of Rameses II...in the central chambers of the great pyramid."*

Elsewhere, during Lister's first Edinburgh years, a young German doctor became interested in germs. Robert Koch had worked as a surgeon in the Franco-Prussian War, where he saw and appreciated the effects of carbolic antisepsis. When he was appointed to a research position, he began the laboratory study of microorganisms and developed techniques for growing bacterial cultures on beds of nutrients ranging from slices of potato to dishes of gelatin. Koch's studies put beyond doubt the germ theory of disease, and founded the science of bacteriology. He became interested in the cause of tuberculosis, at that time widespread and usually fatal, and thought to be an inherited disease. For his discovery of the infective agent, *mycobacterium tuberculosis*, Koch would be awarded the Nobel Prize for Physiology or Medicine in 1905.

While at Edinburgh, Lister was summoned to Balmoral Castle in Aberdeenshire by Sir William Jenner, the Queen's physician, to treat Queen Victoria, who had a large, painful abscess beneath her arm. Lister prepared his carbolic spray machine, his antiseptic dressings, and froze the skin over the abscess with ether to minimise the pain. During the brief excision a slight spray of carbolic acid touched the Queen's face, and she complained to Jenner, who quipped *"Madam, I am only the man who works the bellows!"* The operation was a success, but the wound did not drain completely, and later Victoria was in pain and feverish. Lister took a walk around the grounds of Balmoral to think, and came up with the idea of using a piece of rubber tubing to drain the remaining pus. He went back and cut a length of tube from his spray machine, soaked it overnight in carbolic, and inserted it in the wound the next day. When the dressing was renewed, the wound was clear, and the Queen recovered quickly. Lister incorporated the technique into his practice, and shortly afterwards he was appointed Surgeon to the Queen in Scotland.

At Edinburgh University, Lister's lectures routinely attracted three to four hundred students and his many Infirmary wards were busy with patients. Nevertheless he devoted time to publish and spread knowledge of the effects of antiseptics as widely as possible. The lessons on how to avoid hospitalism had been well absorbed in Copenhagen, and when Lister travelled to German hospitals including Munich, Leipzig, and Berlin, he found that antiseptic practices had been fully adopted, resulting in the practical elimination of gangrene and septicaemia. Adoption of antiseptics in France was also enthusiastic. But uptake of the methods in America and England, especially London, lagged behind. There was jealousy and a feeling that in Lister's wards in the overcrowded Edinburgh Infirmary, he relied too much on antiseptics - he had discontinued the disruptive annual cleaning of wards and seemed proud that *"if we take cleanliness in any other sense than antiseptic cleanliness, my patients have the dirtiest wounds and sores in the world. I often keep dressings on for a week at a time..."* Another surgeon from Edinburgh, Lawson Tait, a one-time student of James Syme but 18 years younger than Lister, favoured aseptic surgery (that is, paying close attention to the hygiene of wounds and the steam sterilisation of instruments) which he had introduced successfully in Birmingham. Lister pointed out in response that the achievement of asepsis in a dirty wound or environment needed the use of antiseptics. His former friend and house-mate, Joseph Sampson Gamgee, was another critic casting doubts northwards from England. The Lancet had changed its tune too, and published a couple of leading articles in 1873 doubting Pasteur, Lister, and the evidence for germs. Lister did not bother to reply.

After eight years at Edinburgh, Lister was tempted south into the lion's den to succeed Sir William Fergusson as professor of clinical surgery at King's College, London. His father-in-law, professor Syme, had passed away in 1870, and since he and Agnes were childless, he had few remaining family ties to Scotland. In London, his appointment was regarded suspiciously, and opinion was that for the first time these experimental antiseptics would be put to a proper test - ignoring the overwhelming evidence from the Continent. Lister brought with him from Edinburgh his two house surgeons, William Watson Cheyne and John Stewart, plus two junior doctors and a small team of nurses, all well-trained in antisepsis. Dr. Stewart gave a detailed account of Lister's first lecture at King's College: *"The subject of the lecture was bacteriology; it was perhaps the first on the subject ever given in London, and as yet the science had not a name. The theme was the actions of germs on milk and putrescible fluids, and he told of experiments he had carried on during the summer...Cheyne and I called on him early in the afternoon. We found him ...busy getting in order the exhibits for his lecture. Mrs. Lister was helping...we drove from his house to the lecture hall in Somerset House."* Lister had chosen his subject because of its relevance to all practitioners of medicine - physicians, obstetricians, surgeons - since, as he said, it covered probably all so-called infectious and contagious diseases. The lecture and the evening welcome dinner that followed were a success. Stewart wrote *"This was a brilliant and most hopeful beginning of what we regarded as a campaign in the enemy's country. But...the next few weeks were to us of his staff the abomination of desolation. There seemed to be colossal apathy, an inconceivable indifference to the light...We four unhappy men wandered about, now in the wards of King's, now through older and more famous hospitals, and wondered why men did not open their eyes. In these wards the air was heavy with the odour of suppuration, the shining eye and the flushed cheek spoke eloquently of surgical fever."* Lister said he used to hang his head when driving to King's at the thought of his handful of apathetic students and his wards of largely empty beds. But in a month or so, suitable patients presented themselves. Lister treated a man with a broken knee-cap by cutting into the bone and wiring it together with silver wire - never done before, and an operation that would have previously been thought reckless though risk of infection and death. A London surgeon said, *"when this poor fellow dies, someone ought to proceed against that man for malpraxis"* [sic]. The patient recovered quickly with a fully restored knee. In another case Lister removed the leg of a patient with a huge cancerous tumour of the thigh, who had been abandoned by other surgeons. King's College staff and students were astonished to find the man sitting up in a day or two reading the newspaper, free of pain or fever. Lister's fellow professor of surgery, John Wood, immediately scrapped the test that had been proposed to compare the results of the two professors, and asked Lister to teach him the antiseptic method. The next converts were Sir James Paget, London's most famous surgeon, and Sir Prescott Hewitt, President of the Royal College of Surgeons. They witnessed Lister's removal of a large tumour on a shoulder-blade of a young lady, at her home, in a case where they both had declined to operate. The necessarily large wound healed perfectly, and the patient had no fever or excess pain. The battle for London minds was won.

Lister continued to innovate and to change his methods if the evidence did not support them. The outcomes of operations done in a cloud of carbolic spray were no better than without it, so he abandoned the practice, and replaced the use of carbolic acid dressings with double cyanide of mercury and zinc, which were less likely to irritate the skin. He held the chair of clinical surgery at King's for fifteen years. During his tenure he was appointed Surgeon Extraordinary to Queen Victoria, and he received a knighthood in 1883, then a peerage in 1897, as Lord Lister of Lyme Regis. In 1902 he received the Order of Merit from Edward VII as one of the original twelve members, who included Lord Kelvin and Lord Rayleigh. Agnes, his devoted wife and research assistant, died in Italy in 1893, and afterwards Lister became increasingly sad and reclusive, though he acted as President of the Royal Society from 1895 to 1900. He died at home in 1912 and his name is honoured and remembered worldwide, including at medical facilities in Glasgow, London and Stevenage, in the bacterium *Listeria*, and in the antiseptic *Listerine*. Albert Einstein, in his last speech in Europe, in 1933, spoke on intellectual freedom. The great physicist said, *"If we want to resist the powers which threaten to suppress intellectual and individual freedom, we must keep clearly before us what is at stake...without such freedom there would have been no Shakespeare, no Goethe, no Newton, no Faraday, no Pasteur and no Lister."*

36 First female doctors of medicine: Elizabeth Garrett Anderson and the Edinburgh Seven (1836-1900)

Elizabeth Garrett's connections with Scotland were, unfortunately, too slight – much slighter than they should have been. In 1862, at the age of 26, she was the first woman to matriculate at the University of St Andrews. Her intention was to continue her studies in the medical sciences, having spent six months as a surgery nurse at the Middlesex Hospital in London. While there she prevailed on the teachers of the hospital's medical school to attend lectures in *materia medica*, chemistry, Latin, and Greek, and gained 'certificates of honour' in each class examination. The outcome made one examiner ask nervously, *"May I entreat you to use every precaution in keeping this a secret from the students?"* In any event, Elizabeth encountered widespread opposition from the male students, who composed a 'memorial' (a memorandum) to the Medical Committee, objecting to Garrett's presence. She was debarred from attending more lectures.

Illustration 43: Elizabeth Garrett Anderson c.1889

Only five students defended her and wrote, *"We, the undersigned students of this Hospital, desire to express our regret…the substance of the memorial is not an expression of the sentiments of the whole of the students of this School of Medicine."* Elizabeth had to leave the Middlesex, and returned to her parents' home to consider her options.

The Garrett family home was in the Suffolk coastal town of Aldeburgh, where Elizabeth's father, Newson Garrett, was a prosperous merchant of coal and barley, and had built an extensive maltings at nearby Snape. After initial opposition, her parents supported Elizabeth's ambitions to qualify as a medical doctor. She was inspired originally by meeting Elizabeth Blackwell, an Englishwoman who was the first female doctor in the United States. Blackwell had visited London in 1859, to lecture on 'Medicine as a Profession for Ladies', invited by the Society for Promoting the Employment of Women, of which Garrett was a member. The feminist Emily Davies was another important early influence. Elizabeth, with her younger sister Millicent, met Davies when Elizabeth was aged 18. Emily would become a life-long friend and correspondent, who later went on to be a co-founder of Girton College, Cambridge. Millicent Garrett would become a prominent campaigner for women's suffrage. One of group's acquaintances was Sophia Jex-Blake, four years younger than Elizabeth, possessor of a stubborn and prickly personality, but who shared their ambitions for the rights of women.

After the disappointment of the Middlesex Hospital, Elizabeth's family encouraged her to continue training in her chosen profession. Cleverly, she used her certificates to gain admission as an apprentice to the Worshipful Society of Apothecaries, whose constitution did not allow discrimination on grounds of sex. Before granting a licence to practice, they would require several years of lectures and examinations at an institution approved by the Society. This was a victory, but despite lobbying led by her father and supported by Mary Somerville, and the Chancellor of the Exchequer, William Ewart Gladstone, Elizabeth was refused entry to the medical schools of Oxford, Cambridge, and

London. She turned her attention to St Andrews, where she knew that the professor of anatomy and medicine, George Edward Day, was prepared to help her. She asked him to approach the University Senate on her behalf and was encouraged by his reply, writing to Emily Davies, *"He seemed much more inclined to be a supporter, [than] when he wrote two months ago, so I suppose he had been talking to some disapproving old fogies."*

At St Andrews, Elizabeth's attendance had the support of Professor Day, the other two medical professors, Heddle and Macdonald, and also the Principal, John Tulloch. She bought a matriculation ticket for the price of £1 that entitled her to attend the relevant classes. The 'disapproving old fogies' were still in control of the University Senate and tried to return her matriculation fee, which she declined to accept. The Garretts took legal advice, threatening a lawsuit, and Newson was prepared to finance one. Emily Davies and Sophia Jex-Blake joined Elizabeth in St Andrews to reinforce her struggle. But in the end, the Senate's old fogies did not budge, and Elizabeth was refused permission to attend formal lectures. Instead, she studied privately with Professor Day. During the following summer of 1863, a similar pattern recurred at the University of Edinburgh, where she was obliged to study privately for a while with James Young Simpson and Alexander Keiller, (who later became President of the Royal College of Physicians of Edinburgh).

Garrett was understandably depressed and frustrated by the serial rejections. She returned to London and was willing to pay for private lessons in anatomy, so that she could pass the examination of the Apothecaries' Hall. She wrote to many medical men to ask for teaching in dissection and surgical anatomy. One reply from 'Mr I.H.' of the University of Aberdeen revealed the mindset of the time. *"I must decline to give you instruction in Anatomy…the entrance of ladies into dissecting-rooms and anatomical theatres is undesirable in every respect, and highly unbecoming…it is not necessary that fair ladies should be brought into contact with such foul scenes – nor would it be for their good any more than for that of their patients if they could succeed in leaving the many spheres of usefulness which God has pointed out for them, in order to force themselves into competition with the lower walks of the medical profession. Ladies would make bad doctors at the best".*

It must have taken a special character to persevere in the face of such blatantly sexist opposition, but Elizabeth continued to look for ways forward. She used her contacts at the London and Middlesex Hospitals to take further informal training, as a 'visitor' rather than a student, until even those permissions were revoked. But by the autumn of 1865 she was ready to take the Society of Apothecaries examination. Their board of examiners had conveniently forgotten their commitment of four years before, and tried to refuse her. Again, her father threatened legal action and issued an ultimatum. Elizabeth was allowed to sit the exam, passed with the top mark, and obtained the diploma licence. Her name was entered on the medical register - the first woman licenced to practice medicine in Britain. The Worshipful Society of Apothecaries swiftly altered their regulations to ensure that future candidates must have studied formally in a recognised medical school – from all of which, of course, women were formally excluded.

Elizabeth began a private practice and then, with funding from her father, founded the St Mary's Dispensary at Seymour Place in London. She also successfully applied for a vacant post on the honorary medical staff of the Shadwell Hospital for Children, impressing the selection board. The board included James G.S. Anderson, a successful, Aberdeenshire-born businessman, and a partner in Anderson, Thomson & Co, a shipping line that contributed to Hospital funds. He was the son of Alexander Anderson, a minister who had joined the dissenters during the Great Disruption of the Church of Scotland, and a nephew of Arthur Anderson, who co-founded P&O Lines, and had been a radical Liberal MP for Orkney and Shetland. James and Elizabeth married in 1871, and they would have three children together – Alan, Margaret and Louisa, who also became a medical doctor, an active feminist, and her mother's biographer. In the few years leading up to becoming Mrs. Garrett Anderson, Elizabeth further expanded her medical knowledge and enrolled for an M.D. degree in the University of Sorbonne, Paris, learning French, studying remotely, and visiting Paris six times to take the examinations. She graduated in 1870, after a *viva voce* examination in French on her thesis, entitled *'La Migraine'*. Even afterwards she continued to meet with resistance, this time from the British Medical Association, who made her a member in 1873 but a few years later passed a motion to exclude women, and maintained that policy until 1892.

Garrett Anderson, already a pioneer, went on to have a phenomenally successful medical career. She expanded the St Mary's Dispensary, renaming it as the New Hospital for Women and Children; became a visiting physician at the East London Hospital for Children; and acted as Dean of the London School of Medicine for Women, created in 1874. The foundation of the London School was a direct result of the experiences of some of Garrett Anderson's friends and colleagues, specifically Jex-Blake, who was the leading figure in a group of women that became known as the 'Edinburgh Seven'.

Sophia Jex-Blake was born in Hastings in 1840, the daughter of Mary Cubitt and Thomas Jex-Blake, a well-to-do lawyer. Sophia had shared Elizabeth Garrett's academic ambitions from a young age, and attended classes in several subjects at Queen's College, London. While coincidentally pursuing some further studies in Edinburgh, she visited Garrett in St Andrews in support of Elizabeth's struggles with the University in 1862. After a spell in Germany teaching mathematics, Jex-Blake spent some time in the United States, after the end of the American Civil War, studying educational methods, and she decided also to pursue a career in medicine. With a fellow aspiring student, Susan Dimock, she published a letter in the *Boston Daily Advertiser* on March 11th, 1867. *"Gentlemen, finding it impossible to obtain elsewhere in New England a thoroughly competent medical education, we hereby request permission to Harvard Medical School asking to be admitted on the same terms ...as other students, there being, as we understand, no university statute to the contrary."* This public request was predictably and publicly refused, with the reply from Harvard stating, *"there is no provision for the education of women in any department of this university."* The two women did not let it rest there. They obtained introductions to the professors of the Medical Faculty, and to each member of staff at the Massachusetts General Hospital, and canvassed them systematically. Many expressions of support were given, only to be overridden by the governing Corporation of the university. Jex-Blake wrote in her diary, *"All which ends…smoke!"* Sophia moved to New York to study and work with Dr. Elizabeth Blackwell, at Blackwell's new hospital, the New York Infirmary for Indigent Women and Children, before returning home to Britain on the death of her father, in November 1868. After a wintertime of grieving and living in Brighton with her mother, she turned again to the pursuit of a medical degree, hoping that after the trailblazing of Elizabeth Garrett, the enlightened 'Scotch' universities would be the most welcoming.

Jex-Blake's chosen target was admission to the University of Edinburgh. She won public backing from *The Scotsman*, Edinburgh's pre-eminent newspaper. She had the friendship and support of Professor Sir James Young Simpson, and David Masson, professor of rhetoric and English literature. Her formal application to the professors of the Medical Faculty to attend summer classes achieved four positive votes, from Allman, Balfour, Bennett, and Simpson. Against were Christison, Laycock and probably Henderson. Absentees included the opponent James Syme and the supportive Lyon Playfair. The question passed next to the Senate, composed of all the University's professors, who voted 14-4 in favour. Success was not assured. Appeals were made to the University Court, which consisted of the Rector, the Principal, the Lord Provost of Edinburgh, and five representatives appointed respectively by the Rector, the Senate, the Town Council of Edinburgh and the General Council of graduates of the University. Unfortunately for Jex-Blake, the Senate representative was the widely-respected Professor Robert Christison, President of the Royal Society of Edinburgh, soon to be knighted and the only medical man and the only professor who had a seat on the University Court, and who was implacably opposed to the admission of women to medical studies. On April 19th 1869, the decision was *"That the Court, considering the difficulties at present standing in the way of carrying out the resolution of the Senatus, as a temporary arrangement in the interest of one lady, and not being prepared to adjudicate finally on the question whether women should be educated in the medical classes of the University, sustains the appeals and recalls the decision of the Senatus."*

The bad news was delivered by Professor Masson who wrote

Illustration 44: Sophia Jex-Blake, leader of the Edinburgh Seven

to Jex-Blake, suggesting that *"if a new attempt were to be made…in the form of a joint and simultaneous application from a few ladies (say from half a dozen to a dozen), then our authorities would be obliged to yield"*. The campaign in *The Scotsman* had achieved widespread publicity and given hope too. Jex-Blake advertised in the newspaper for like-minded women to join her, and a month after the Senate decision, she received a letter from a 34-year-old woman in London, Isabel Thorne, who wanted to join the fray, writing, *"I should be glad, if you renew your application, to join you in doing so, and I believe I know two or three other ladies who would be willing to do the same"*. Within a few months, a group of women had formed who were intent on achieving medical degrees. Edith Pechey, aged 23, had worked as a medical apprentice with Elizabeth Garrett, but was doubtful of her own academic abilities, especially in Latin and 'Euclid'. Matilda Chaplin, just turned 23, was the youngest of the group and had already passed the preliminary exam of the Society of Apothecaries before it effectively bolted its doors against women. Helen Evans was a recent military widow, aged 35. Mary Anderson, 32, from Banffshire, was the only

Scotswoman of the group. Emily Bovell, 28, had been a tutor of mathematics at Queens College, London.

In the summer of 1869, the seven women applied to matriculate at the University of Edinburgh and their applications were approved. There was an entrance examination, in two parts: initially, mathematics, English and Latin; then two subjects from Greek, French, German, natural philosophy, logic and moral philosophy, and higher mathematics. In October, 152 candidates sat for the exam, including five of the 'Seven'. All passed, and four of them came in the top seven places. They became the first women to have their matriculation confirmed at any British university. Mary Anderson and Emily Bovell would successfully take the exam the next autumn, with 'Miss Bovell's French best in University except for one Frenchman's.' Meanwhile Edinburgh University had been working on a plan. There would be separate medical classes for women, to avoid the possible embarrassment of 'indelicate' subjects being taught to a mixed audience. The women's fees were set higher, due to the smaller size of their classes.

Settled in Edinburgh, their studies went well over the winter of 1869-70. In the spring, the first exams in chemistry and physiology were set. Edith Pechey, despite her self-doubts, came in first place of all students sitting the exams for the first time. This provoked the first sign of trouble. Robert Christison's hostility to women students was undimmed. He awarded 'Hope Scholarships' for top exam results not to Pechey, but to men with lower marks. Ironic, since the Scholarships had been founded by the late Professor Hope 50 years before, from the proceeds of 'lectures given to ladies'. Christison then successfully opposed a University Court motion to amalgamate the male and female classes, and used his influence with other members of the medical faculty to turn them against the women. Some male students became hostile too, and the women suffered harassment, foul verbal abuse, and the nameplates and doorbells of their lodgings were vandalised. The summer break must have been welcome.

Admission to the next session's classes in anatomy and clinical studies proved to be what Jex-Blake called a 'Giant Difficulty', but after many arguments the women were admitted to the mixed-sex anatomy classes and to the Dissection Room at Surgeon's Hall. Again, they distinguished themselves. Never, reported the lecturer, Dr. Handyside, had better work been done in his classrooms. The next battle was against the refusal of the managers of Edinburgh Royal Infirmary to give the women access to practical training in their wards. Handyside stepped in to point out that they were all fully accredited medical students of the University. The 'Seven' co-signed a letter to the Infirmary citing five of their own doctors who were willing to mentor them. But resistance among the University student body had gathered momentum - a petition against the women's admission to the Infirmary gained 500 signatures. On Friday the 17th of November 1870, the women approached the Surgeon's Hall for a demonstration and examination in anatomy. They found their way barred by a mob of nearly 200 students, jostling, shouting abuse and throwing mud. As the women got to the entrance gate, it was clashed shut in their faces, by some students inside who were blowing smoke, cursing and drinking from bottles of whisky. Shouts of "shame' and 'let them in' came from some others, until the gate was opened by the janitor. The anatomy lecture room was crowded to the door, until Dr. Handyside ejected those students not belonging to his class. Someone then put a sheep into the room, causing Handyside to say, *"Let it remain – it has more sense than those who sent it here."* Three students were later found guilty of a breach of the peace and fined £1 each. What became known as the 'Surgeon's Hall Riot' shocked and energised support from some male students, including a chivalrous 'Irish Brigade', who began to act as bodyguards, and arranged to escort the women to and from lectures and examinations. Jex-Blake accused Professor Christison's classroom assistant of fomenting the riot, and was in turn sued for libel. In a trial brought before a hugely antagonistic judge, Lord Mure, who refused to allow the class-assistant to be questioned, the jury was directed to find in favour of Christison's assistant but awarded damages of just one farthing, having been assured that Jex-Blake would not be made liable for the costs of the trial. In fact, Lord Mure did exactly that, and made her liable for the costs of £915-11s-1d. They were paid by public subscription following the placement of a newspaper appeal, and the formation of a 'Committee for securing complete Medical Education for Women in Edinburgh'.

The women battled on in the face of persistent opposition from the Medical Faculty, led by Robert Christison, who tried to deny them access to examinations and even to matriculate for the following session. Some battles were won, but a war of attrition was being fought. In November 1871, the Senate voted 14 to 13 to rescind the regulations in favour of female students and to take no steps to enable them to complete their education, only for the decision to be overturned by the University Court. Most of the Royal Infirmary doctors continued to refuse female students admittance to their wards (despite, of course, the ubiquitous presence of female nurses). By the end of 1871, three of the original seven women had, understandably, had enough. Helen Evans withdrew to marry Alexander Russel, editor of *The Scotsman*. Matilda Chaplin gave up and married a physicist, William Ayrton, and they emigrated to Japan. Mary Anderson married a solicitor from Greenock, (who tragically died within two months of their wedding, and her infant son by him died the next year). The remainder of the 'Seven' persevered, together with several other women who had enrolled as medical students, and with growing support from the public. But they were to suffer yet more persecution at the hands of their University opponents.

In early 1872, the University Court, again persuaded by the dogged Professor Christison, pronounced that they could not offer degrees; but if the women 'gave up the question of graduation', they would try to provide 'certificates of proficiency' – quite useless to allow entry to medical practice. Jex-Blake and her colleagues, with the apparent support of the Lord Advocate of Scotland, brought a legal action against the University Senate to confirm their right to graduate. Six professors, including Masson, refused to defend the legal action, and recorded an eloquent statement of support for the women's rights. Only one medical professor, Bennett, was among them, though Fleeming Jenkin asked for his name to be removed from the list defending the action. A preliminary judgement found in favour of the women; an appeal by the University to the Court of Session found against them by seven judicial votes to five, on the faintly ridiculous grounds that the University's original decision to admit the women had been *ultra vires* – beyond its powers. The legal costs were thrown on the students. They considered an appeal to the House of Lords, but decided, given the bitter animosity of most of Edinburgh's Medical Faculty, to focus their energies elsewhere.

Jex-Blake, with another five of the Seven, came together to found the London School of Medicine for Women, in 1874, and most of the students who initially enrolled, transferred from Edinburgh. Elizabeth Garrett Anderson agreed to serve as Dean of the School, but thought that Jex-Blake was temperamentally unsuited to be in charge, and appointed Isabel Thorne to that role. Jex-Blake and Edith Pechey gained their M.D. degrees in Bern, and Emily Bovell gained hers in Paris, all in 1877. Both Matilda Chaplin Ayrton, returned from Japan, and Mary Anderson followed Bovell in graduating from Paris two years later, and they subsequently practised medicine in Garrett Anderson's New Hospital for Women in London. Isabel Thorne never qualified as a doctor, but in her role as honorary secretary of the London School of Medicine for Women, used her managerial skills to solidify its establishment. Helen Evans was widowed in 1876 when Alexander Russel died suddenly of heart failure, leaving her with three children. She went on to be a member of school boards in Edinburgh and St Andrews. Jex-Blake returned to Edinburgh in 1878 and established a medical practice, and later the Edinburgh School of Medicine for Women, joined on the executive board by Evans. Their vehement opponent, Sir Robert Christison, continued in prominent medical practice for many years, and died at home in Edinburgh in 1882, aged

84. Elizabeth Garrett Anderson retired to Aldeburgh, where in 1908 she became the first woman mayor in Britain, and she continued to fight for the cause of women's suffrage and liberation. The Scottish universities finally, and completely, opened their doors to women, in 1892. The Edinburgh School of Medicine for Women was plagued by financial problems and personality clashes. It lost students to London, Dublin and Glasgow, and eventually closed in 1898. Sophia Jex-Blake retired to live in Sussex with her partner, Dr. Margaret Todd, who would be her biographer.

The pioneering bravery and battling spirit of Elizabeth Garrett Anderson and the Edinburgh Seven had a profound impact and a lasting legacy. The Elizabeth Garrett Anderson wings at Ipswich Hospital, and University College Hospital, London are just two examples of her commemoration. The Edinburgh Seven were awarded posthumous honorary medical doctorates by the University of Edinburgh in July 2019. The degrees were received on their behalf by a group of women students from the Edinburgh Medical School. Perhaps just as poignantly, in 1895, Agnes Forbes Blackadder became the first female graduate of the University of St Andrews, and gained her medical degrees from Glasgow University in 1901. But that is another story.

37 Magneto- and electro-optics: John Kerr (1824-1907)

Illustration 45: John Kerr, physicist

In the introduction to his first scientific paper, in the *Philosophical Magazine*, John Kerr wrote in 1877, *"I was led some time ago to think it very likely, that if a beam of plane-polarized light were reflected under proper conditions from the surface of intensely magnetized iron, it would have its plane of polarization turned through a sensible angle in the process of reflection."* This was not an obvious thought at a time when magnetism was still poorly understood. Kerr's thoughts had been prompted by the experiments of Michael Faraday, around 30 years earlier, which had discovered what is now called the Faraday Effect, the rotation of the plane of polarization of light passing through lead glass (or a similar medium) within a magnetic field, which was the first evidence that light and magnetism are somehow related. Kerr had also pondered the implications of Brewster's Law (which showed the deep relationship between the reflected and refracted parts of light rays at the surfaces of optical media) and the known laws of metallic reflection, whereby the intensity of reflected light has a dependence on the polarization angle of the incident light.

Kerr tested his idea in an experimental set-up with a mirrored surface, which was the polished end of an iron cylinder that formed the core of a horseshoe electro-magnet. Using light from a paraffin flame polarized using a Nicol prism, the standard polarizer of the day, he reflected the polarized light from the end face of the cylinder at various settings of the magnetising current, and analysed the reflected light with another 'Nicol'. The finding partly proved his suspicion. The polarization of the reflected light was altered by the magnetic field, but not quite in the way he anticipated. He found that *"the beam reflected by the mirror of magnetized iron is certainly not plane-polarized, as is the incident beam (and the reflected beam also before magnetization); for when the light is restored by magnetic force from pure extinction ... it cannot be extinguished by any rotation of the second Nicol in either direction."* Actually, the reflected light was elliptically polarized, that is, it consisted of a mix of 'horizontal' and 'vertical' polarizations, out of phase with each other, and oriented at an angle dependent on the angle of incidence and the strength of the magnetic field in the metal. This effect, now known as the magneto-optic Kerr Effect (MOKE) was in fact the second useful optical effect discovered and named after John Kerr. His first discovery was a change in the refractive index of materials subjected to an electric field, proportional to the square of the electric field's strength, and first described by Kerr in the same *Philosophical Magazine* in 1875. He found the effect firstly in glass and other solid dielectric (insulating) materials and then in liquids. In all materials, this change in the refractive index occurs, more strongly in some materials than in others. Specifically, the application of an electric field causes the material to become birefringent, with an index of refraction that is different for light polarized parallel or perpendicular to the electric field. Again Faraday had speculated that such an effect might exist, but he could not detect it with his available equipment.

At the time of his important experiments, which would lead to many practical applications, John Kerr was a lecturer in mathematics at the Free Church Training College in Cowcaddens Street, Glasgow. The college, one of the forerunners of Jordanhill College of Education, was established in 1845 soon after the Disruption in the Church of Scotland required Free Church adherents to relinquish state-funded teaching posts. Kerr, educated in a village school in Skye and then at the University of Glasgow, was an ordained minister in the Free Church, but never practiced actively. At Glasgow he had initially studied divinity, but then, under the tutelage of Professor William Thomson (later Lord Kelvin and only six months older than Kerr) he graduated as the top student in mathematics and natural philosophy, winning the Earl of Eglinton Prize of 20 sovereigns. Kerr and Thomson became life-long friends and Kerr helped Thomson to set up his physical laboratory, the first in a British university. Kerr's own small laboratory at the Free Church Training College was mainly equipped at his own expense and used in his spare time.

Kerr could not have anticipated the important practical applications of his discoveries. Magneto-optic Kerr Spectroscopy is used to examine the material properties of metals and rare-earth alloys, and has become an important tool for controlling the production quality of magnetic recording media. The electro-optic Kerr Effect is utilised in the Kerr Cell, typically a small glass cell containing a liquid such as nitrobenzene or KTN (potassium tantalate niobate). When a strong electric field is applied across the cell, the induced birefringence causes plane-polarized light to be split into two components, parallel and perpendicular to the direction of the field. The two components travel at different speeds through the cell, resulting in a difference in phase between them so that the emerging light is circularly or elliptically polarised, and so has a component that can be transmitted by an analysing polarizer crossed to the incident plane. Because of the speed of the effect, Kerr Cells can be used as high-speed optical shutters allowing shutter speeds down to around 100 nanoseconds. Early Kerr Cells were used to modulate light in television receivers, including the early Baird televisions in the 1930s. Kerr Cell shutters have been used to measure the speed of light, and can be used within the optical cavities of lasers to produce high-power, short pulses in techniques known as 'Q switching' and 'mode-locking'. In optical fibres transmitting high intensity light, the electric field of the light itself can induce the electro-optic Kerr Effect, causing modulation of the fibre's refractive index and producing 'solitons', the remarkable self-sustaining wave pulses first observed in their watery form on the Edinburgh-Glasgow Union Canal by John Scott Russell.

Apart from his seminal publications on magneto- and electro-optics, John Kerr championed the adoption of the metric system in Britain and wrote an influential physics textbook *'An Elementary Treatise on Rational Mechanics'*, published in 1867. His contributions as an educator were recognised by the award of an honorary Doctorate the next year, by the University of Glasgow, where his apparatus and early Kerr Cells are still preserved. He was elected a fellow of the Royal Society in 1890 and received its Royal Medal in 1898. Kerr retired from his teaching post in 1901 and was awarded a civil list pension of £100 per annum. With his wife Marion, he had three sons and four daughters. He died in Glasgow, aged 83 in 1907, just a few months before Lord Kelvin, his greatest friend and colleague.

38 The Rankine Cycle: William John Macquorn Rankine (1820-1872)

Illustration 46: William John Macquorn Rankine, scientist and engineer

William J. M. Rankine was one of the founders, together with Rudolf Clausius and William Thomson (Lord Kelvin) of the science of thermodynamics. Moreover, he applied the science to practical issues and developed the theory underpinning the operation of steam engines - indeed of all heat engines, pointing out the extension needed to Carnot's theory. His work on the thermodynamics of steam engines led to the Rankine Cycle, used to generate electricity in modern power plants.

Rankine was born in Edinburgh, the son of David, (a railway engineer who claimed descent from Robert the Bruce) and his wife Barbara Graham, daughter of a Glasgow banker, and second cousin to the famous chemist Thomas Graham. He studied at Ayr Academy and briefly the High School of Glasgow, but he had prolonged periods of illness that required him to be educated privately too. He matriculated at the University of Edinburgh in 1836 but without registering for a degree, instead taking a wide range of classes in natural philosophy, botany and natural history, with extra-mural studies in chemistry and philosophy. The professor of natural philosophy, James David Forbes, awarded Rankine two gold medals for essays on 'the wave nature of light' and 'methods of physical investigation'. Rankine resolved to become an engineer, and spent summers working for his father on the Dalkeith railway scheme before taking an apprenticeship with the civil engineer Sir John Macneill in Ireland. While working on surveys of the Dublin and Drogheda Railway, he devised the geometrical rules of setting out circular curves, called Rankine's Method. He became an associate of the Institution of Civil Engineers while working on projects for the Caledonian Railways. Meanwhile he had developed an interest in music and poetry, and in 1842 he became a fellow of the Royal Scottish Society of Arts.

Rankine's early publications included a pamphlet called *'An experimental enquiry into the advantages of cylindrical wheels on railways'* and a paper on the *'Fracture of axles'* where he identified the gradual progress of metal fatigue and disproved the then-current theory of the spontaneous repair of cracks through re-crystallisation of the metal. His thoughts quickly turned to the nature of heat, which at that time was assumed to be a material substance - caloric - which could flow from one body to another. Carnot's 1820 theory of the heat engine, although largely correct, also made this assumption. Rankine thought otherwise. By considering the effect of heat of the pressure and volume of gases, he concluded that heat must be a form of molecular motion, which he called *molecular vortices*, and worked out the theory in quite brilliant mathematical detail. His theory supposed that solid bodies and gaseous vapours, apparently at rest, are composed of molecules in motion, and that the heat content of the substance is just exactly the energy of the molecular motion. He imagined the molecules to be whirling about their axes, writing *"the hypothesis of molecular vortices may be defined to be that which assumes that each atom of matter consists of a nucleus or central point enveloped by an elastic atmosphere, which is retained in its position by attractive forces, and that the elasticity due to heat arises from the centrifugal forces of those atmospheres, revolving or oscillating about their...central points."* We now know that the motion of molecules in gases is a mixture of high-speed linear, vibrational and rotational motion, but Rankine's model captured the essence of the truth. William Thomson, professor of natural philosophy at Glasgow, presented a paper in 1849 to the Royal Society of Edinburgh asking *"how must the theory of the heat engine be modified, supposing that heat is not a substance, but a mode of motion?"* Rankine already had the answers, soon published in two seminal papers in 1850 called *'On the mechanical action of heat, especially in gases and vapours'* and *'The centrifugal theory of elasticity as applied to gases and vapours'*. These used his theory of molecular vortices to deduce the links between temperature, density, and pressure of vapours, gases, and liquids and showed close agreement with new experimental data. Rankine also showed where Carnot was wrong: it is not heat that is conserved in a heat engine as some kind of indestructible 'caloric', but the sum of heat and the work produced. In other, modern, words it is *energy* that is conserved in the system. He elaborated Carnot's correct deduction that the efficiency of a 'heat engine' depends solely on the temperatures of the heat source (T_1) and the waste heat 'sink' (T_0) - Carnot had not detailed the relationship. Rankine derived the exact dependence of the ideal engine efficiency on the two temperatures. He showed that the efficiency is equal to the *absolute temperature difference divided by the absolute temperature of the heat source*, that is:

Ideal efficiency = ratio of useful work produced to heat input
$$= (T_1-T_0)/T_1$$

Clearly, the efficiency can be increased to approach 100% either by reducing the lower temperature as close as possible to absolute zero, or by increasing the higher temperature as far as material properties will allow.

Furthermore, Rankine proposed a practical power cycle that maximises the efficiency of real steam engines. He did this by devising a closed-cycle two-phase process, which uses water and steam in a closed loop, rather than exhausting waste steam. The Rankine Cycle is used in modern power plants (whether coal, gas, or nuclear-powered) to generate electricity and goes like this:

Step 1: the water is compressed from low to high pressure; this raises its boiling point, enabling high-temperature steam to be produced in the next step

Step 2: the high-pressure water is heated at constant pressure in a boiler, turning it to 'dry' saturated steam, i.e. steam devoid of non-vaporised water droplets

Step 3: the high-pressure saturated steam is allowed to expand through a turbine, producing power and electricity; the steam loses pressure and temperature, and experiences some condensation

Step 4: the 'wet' steam is introduced to a low-temperature condenser, turning it back to liquid water.

So, the working fluids in the Rankine cycle (water and steam in this example) are constantly re-used in a closed loop. The steam that is often seen emitting from modern power stations comes from the cooling towers used to condense the working fluid, not from the steam driving the turbines.

In 1853 Rankine was elected a fellow of the Royal Society and two years later he was appointed, aged 35, to the relatively new Regius chair of Civil Engineering and Mechanics at the University of Glasgow, endowed and paid for by the government. He succeeded the inaugural Regius professor, his friend Lewis Gordon. Regius professors at Glasgow University sat outside the controlling, protective, and instinctively conservative Faculty which consisted of the Principal and 13 professors. On Gordon's resignation to pursue his lucrative engineering practice, the faculty attempted unsuccessfully to redirect the Regius endowment. After his appointment Rankine allied himself to the reforming Whig faction in the University and proved to be a highly productive researcher and popular teacher - Rankine was a character. His friend and colleague professor of physics, Peter Guthrie Tait, wrote *"His appearance was striking and prepossessing in the extreme, and his courtesy resembled almost that of a gentleman of the old school. His musical tastes were highly cultivated, and it was always exceedingly pleasant to see him take his seat at the piano to accompany himself as he sang some humorous or grotesquely plaintive song..."* One of Rankine's jokey compositions parodied the views of the British workman on metric units:

When I was bound apprentice, and learned to use my hands,
Folk never talked of measures that came from foreign lands:
Now I'm a British Workman, too old to go to school;
So whether the chisel or file I hold, I'll stick to my three-foot rule

Some talk of millimetres, and some of kilogrammes,
And some of decilitres, to measure beer and drams;
But I'm a British Workman, too old to go to school,
So by pounds I'll eat, and by quarts I'll drink, and I'll work by my three-foot rule

A party of astronomers went measuring the earth,
And forty million metres they took to be its girth;
Five hundred million inches, though, go through from pole to pole;
So let's stick to inches, feet and yards, and the good old three-foot rule

During Rankine's tenure at Glasgow he was prodigiously productive, publishing several hundred papers and textbooks, not just on civil engineering and mechanics but also on thermodynamics, ship design, materials science, heat engines and fluid mechanics. He is especially revered as a co-founder of the science of thermodynamics, with William Thomson and Rudolf Clausius. James Clerk Maxwell wrote, in *Nature* in 1868, *"Of the three founders of theoretical thermodynamics, (Rankine, Thomson and Clausius), Rankine availed himself to the greatest extent of the scientific use of his imagination. His imagination, however, though amply luxuriant, was strictly scientific."* He defined a scale of absolute temperature (in degrees Fahrenheit) now known as the Rankine Scale, and a 'Rankine Body' is an important concept used to compute the flow of fluid around a solid object or surface. As an engineer, businessman and administrator Rankine worked with the Glasgow Philosophical Society to revive and execute the plan to bring water to the city from Loch Katrine; co-founded the 'Institution of Engineers and Shipbuilders in Scotland'; and was active in raising funds for the new Glasgow University buildings in Kelvingrove. He tirelessly promoted the concept of 'engineering science' and successfully pressed for the creation of a B.Sc. in Engineering, which was established in 1872. The same year Rankine died, unmarried, with failing sight and mobility, at home in Glasgow, only 52 years old. He was succeeded as professor of civil engineering by James Thomson, elder brother of William Thomson, Lord Kelvin. The University and the city mourned the passing of a true and brilliant polymath - mathematician, engineer, natural philosopher, businessman, teacher, administrator and musician - whose rigorous development of his insights co-created the new science of thermodynamics, and energised commerce and industry in the second city of the empire.

39 Absolute temperature and Thermodynamics: William Thomson - Lord Kelvin (1824-1907)

Illustration 47: William Thomson, aged 22 – by John Graham Gilbert (Glasgow Museums Resource Centre)

At the age of 22, William Thomson was appointed as professor of natural philosophy in the University of Glasgow, a Chair established in 1727 in one of the oldest universities in Great Britain. Just a few years before his appointment, he had been a first-year student in the University. Thomson was a remarkably intelligent young man, with enormous academic ability. His background was fairly remarkable too. He was born in Belfast, the second son among the seven children of Margaret Gardner and James Thomson. The Thomsons had originally migrated to Ulster from Scotland in around 1640 during the turbulent years before the English Civil War and the resulting Wars of the Three Nations. William's mother died when he was only six years old. His father James had been a farm labourer as a young man, who taught himself mathematics and astronomy, and through sheer dedication, while working as a teacher in Ulster during the summers, managed to study at Glasgow University where he graduated with an M.A. in 1812. After some years teaching mathematics in Belfast, ultimately as professor at the Royal Belfast Academical Institution, James was appointed to the chair of mathematics at Glasgow, in 1832, two years after the untimely death of his wife. He started work, and then moved his bereaved family of young children - aged three to twelve years old - to Glasgow. He kept the education of William and William's older brother, also named James, in his own hands, while their two younger brothers were taught by their elder sisters. During the six-month winter sessions at the University the family lived in the old Glasgow College buildings on the High Street, and in the summer months decamped to rented places on the Firth of Clyde, including Largs, and Lamlash on the island of Arran.

William and James informally attended their father's University lectures, William from the age of ten. James was two years older. They would repeat at home some of the practical demonstrations they saw, making electrical batteries, machines, and Leyden jars to give electric shocks to their friends. In October 1834 they both officially enrolled as students, when James was twelve years old and William ten years and three months. William initially attended lectures in natural history and Greek, and won prizes in both subjects. The next year William and James won the first and second prizes in the junior mathematical class, and then proceeded to 'duopolise' all the honours in the senior mathematics class, and in logic, astronomy and natural philosophy. William later recalled his years at the University with great fondness, praising the breadth of his education and remembering *"the little tinkling bell in the top of the college tower, calling college servants and workmen to work at six in the morning; the majestic tolling of the great bell wakening at seven the professors (and students, too, in the olden times when students lived in the college)."*
William studied his physics initially under Professor William Meikleham, and then the professor of astronomy, John Pringle Nichol, who stood in when Meikleham's health declined. Nichol introduced William to Joseph Fourier's *'Theorie analytique de la chaleur'* (analytic theory of heat)

which would influence Thomson's whole career. He recalled *"the origin of my devotion to these problems is that after I had attended in 1839 Nichol's senior Natural Philosophy class, I had become filled with the utmost admiration for the splendour and poetry of Fourier... I took Fourier out of the University Library; and in a fortnight I had mastered it - gone right through it."* Fourier's book was the first mathematical treatment of heat diffusion, but the mathematics it contained (now called Fourier Series and Integrals) had much wider applicability. Such 'Continental' mathematics was disparaged or ignored by British mathematicians still in thrall to SIr Isaac Newton's work - particularly Philip Kelland, the then professor of mathematics at Edinburgh. Their criticisms prompted Thomson (using the pseudonym P.Q.R.) to defend Fourier in his first two scientific papers, published in the *Cambridge Mathematical Journal*. In a third paper he also showed how Fourier analysis could be used to model the flow of electricity in conductors. Absorbing Fourier's book so quickly had required fluency not just in mathematics but in French, which William had learned well during a two-month stay in Paris over the summer of 1839. But not content with his offsprings' linguistic capabilities, their father arranged two months of lessons in German conversation for the whole family, and then took Elizabeth (aged 22), James (18), William (16) and Robert (11) to Germany for all of the next summer. It was while in Frankfurt that William first read Kelland's criticism of Fourier, and said later, *"My father took us to Germany and insisted that all work should be left behind, so that the whole of our time should be given to learning German...Now, just two days before leaving Glasgow, I had got Kelland's book (Theory of Heat, 1837) and was shocked to be told that Fourier was mostly wrong. So I put Fourier in my box, and used in Frankfort* [sic] *to go down to the cellar surreptitiously every day to read a bit of Fourier. When my father discovered it he was not very severe upon me."* William immediately saw the mis-understandings in Kelland's criticism and on the spot wrote the first 'P.Q.R.' paper, correcting Kelland's errors. When received by the Cambridge journal, the editor, Duncan F. Gregory, sent it for comment to Kelland, who was understandably piqued. A judicious re-write by William and his father to smooth the wording of a few passages ensured that it was accepted and published in 1841, with Kelland now agreeable and *"charmed with the paper"*.

In 1841 William Thomson left Glasgow for the University of Cambridge. Although he had (of course) passed his final exams at Glasgow, he did not formally accept his degree in order to ensure that he could enrol as an undergraduate in Cambridge, which he did, at St. Peter's College - Peterhouse - on April 6th, a couple of months before his 17th birthday. Thomson's Cambridge years were both happy and successful. He became an enthusiastic daily swimmer and rower, won sculling competitions, began to play the cornet, and continued his impressively young contributions to science and mathematics. In his first term the first two P.Q.R. papers appeared and the identity of the author soon leaked out, marking him as a future contender for 'Senior Wrangler' - the top mathematics graduate in Cambridge, once described pretentiously as 'the greatest intellectual achievement attainable in Britain' (and achieved seven years previously by a certain Philip Kelland). William and his father wrote to each other frequently, on the various subjects of William's life in Cambridge. Academic progress, expenses, and time spent rowing were frequent topics of animated discussion. William's purchase of a half-share in a one-man rowing boat at the cost of seven pounds caused his father to express surprise and cast doubt on the wisdom and value of the expenditure. William subsequently won the prestigious 'Colquhoun Silver Sculls' race and continued to distinguish himself academically by winning the Gisborne scholarship award of £30 a year. One letter from father to son in early 1843 conveyed a nascent, secret thought. William Meikleham, still professor of natural philosophy at Glasgow, was in poor health and ailing. James was already forming the ambition that William should succeed Meikleham to the Chair. The possible rivals included D.F. Gregory (though it was said *"a mere mathematician would not be able to keep up the class"*) and the formidable James Forbes, professor of natural philosophy at Edinburgh. In discreet conversations with his university colleagues about his ambition for his son, Professor Thomson found that William's extreme youth might not be an obstacle, but his lack of a background in experimental science, including chemistry, might well be. He at once exhorted William to contact the professor of chemistry at Cambridge to rectify the gap. William compromised by agreeing to acquire some apparatus to study aspects of the polarization of light, and by spending a summer month in the University of Glasgow's chemistry lab.

Thomson's final Cambridge examinations took place in the first days of January 1845, after which he wrote another original paper (this time in his own name) for the *Cambridge Mathematical Journal*, titled *'Demonstration of a Fundamental Proposition in the Mechanical Theory of Electricity'*, in which he showed that for any charged particle in equilibrium between a number of forces obeying the inverse-square law, the equilibrium is unstable. Thomson's exam results gave him the position of Second Wrangler, beaten, to general amazement, by Stephen Parkinson, a man from St. John's College who would have a mathematical career of no particular originality. But Thomson won the top Smith's Prize for advanced mathematics, and in the Tripos degree exams one of his distinguished mathematician examiners remarked to another *"you and I are just about fit to mend his pens."* William Thomson's education was formally complete, and his career as one of the most productive and pioneering physicists of all time was already under way. The cost of his three-and-a-bit-year Cambridge education, funded in small part by scholarship prizes but almost entirely by his anxious father, amounted in his father's reckoning to be a surprisingly large £774-6s-7d. Professor Thomson wrote to his son. *"How is this to be accounted for? Have you lost money or been defrauded of it, or have you lived on a more expensive scale? Do consider the matter and state fully and clearly what you take to be the reason..."*

Expense notwithstanding, in 1845 William spent some months in Paris, accompanied by his best Cambridge friend, Hugh Blackburn, who would later succeed William's father as professor of mathematics at Glasgow. It was work as well as play. Thomson had letters of introduction from Sir David Brewster to Arago, Biot and Babinet, from Philip Kelland to Cauchy, and from his Cambridge tutor William Hopkins, to the brilliant mathematician Joseph Liouville, whose work Thomson greatly admired. In turn Biot introduced Thomson to Victor Regnault, professor of natural philosophy at the Collège de France. Regnault was a superb experimentalist, and was researching the laws governing the performance of steam engines. Generously, Regnault allowed Thomson to assist in his laboratory experiments on the density of gases, which helped to fill the gap in William's practical experience. French sojourn complete, Thomson returned to Peterhouse as Foundation Fellow and to work as a college lecturer, private tutor, and assistant examiner. He became editor of the renamed *Cambridge and Dublin Mathematical Journal* following Duncan Gregory (who sadly had died at just 30 years old the previous year) and published several papers on the mathematical theory of electrical forces, while corresponding with Liouville and Michael Faraday.

In May 1846 the chair of natural philosophy at Glasgow became vacant on the death of William Meikleham, who had held it since 1803, but whose duties had been partly fulfilled in his last years by John Nichol, the astronomy professor who had introduced Thomson to the work of Fourier. Thomson applied for the post and provided glowing testimonials from (among others) William Hopkins; the Masters and Fellows of St. Peter's College, Cambridge; James Forbes of Edinburgh University; Joseph Liouville; and Victor Regnault. The testimonials in support of his son were proudly and assiduously collected by the *in-situ* Professor Thomson and politely circulated to the members of the Glasgow faculty. One of them, Professor Maconochie, wrote back humorously, *"It is plain from the testimonials that your son is, or is shortly to be, a 2nd Newton. He finds a warm advocate in Mrs. Maconochie, but I am much afraid of a father and son in our Faculty meetings. Pray what <u>politics</u> does your William profess?... the question has been put to me, to which I can return no answer further than saying that I believe his father to be a pestilent Whig...Is he a dutiful son, or will he sometimes oppose his father, who in my opinion is generally wrong?*

Reply at your leisure." Apart from this whimsical reservation there were more serious concerns that William's lectures would be too mathematically advanced and 'above' the Glasgow students. Nevertheless Thomson was elected to the post unanimously, ahead of his strongest competitor, David Gray, professor of natural philosophy at Marischal College, University of Aberdeen. His father's reaction was recorded by William's brother-in-law, Dr. David King, writing to his wife, Elizabeth. *"The first announcement I had on the subject was your father's face as he came out of the hall where the election was conducted. A face more expressive of delight was never witnessed. The emotion was so marked and strong that I only fear it may have done him injury."* William took up residence in the University premises at No. 2 The College, living with his father and his widowed aunt, Mrs Gall, who kept house for them.

William Thomson's professorial challenges were immediate. The post had been partly neglected over the seven years since he himself had attended classes as an undergraduate. He had quickly to establish his authority over a class of 100 students, and build credibility within the conservative faculty of the University, presided over by the 75-year-old Principal Macfarlan, who opposed all academic and political reform. Thomson succeeded in both, and won financial support for the modernisation of the physical apparatus in his department, which saw the latest electrical, optical and acoustic instruments installed. His enthusiastic lectures were enjoyed by his students, most of whom were of his own age, though his passion for mathematical analysis was indeed sometimes beyond them. One said, *"I listened to the lectures on the pendulum for a month, and all I know about the pendulum yet is that it wags."* The new professor rediscovered a long-forgotten model of Robert Stirling's highly efficient heat engine, donated by Stirling 40 years before. This reinvigorated his interest in the Carnot-Clapeyron theory of the motive power of heat, which he had encountered in France, where he had sought but failed to find a copy of Carnot's original but little-known work, published in Paris in 1824 as *'Réflexions sur la puissance motrice du feu et sur les machines propres à développer cette puissance'* (Reflections on the motive power of heat and on machines fit to develop that power*).*

Carnot's stated objective had been *"to establish principles applicable not only to steam engines but to all imaginable heat-engines, whatever the working substance and whatever the method by which it is operated."* Although he regarded heat to be the indestructible fluid 'caloric', Carnot had reached some insightful conclusions: *"Wherever there is a difference in temperature, wherever it is possible for the equilibrium of the caloric to be re-established, it is possible to have also the production of impelling power. Steam is a means of realising this power, but it is not the only one."* He established the principle of what is now known as the Carnot Cycle, the ideal relationship between the temperature and volume of the working substance of any heat engine: *"Heat can evidently be a cause of motion only by virtue of changes in volume or of form which it produces in bodies..."* and he considered the maximum efficiency with which that can be achieved: *"the necessary condition of the maximum is, then, that in the bodies employed to realise the motive power of heat there should not occur any change of temperature which may not be due to a change of volume. Reciprocally, every time that this condition is fulfilled the maximum will be attained."* Thomson had in the end got his hands on a copy of Carnot's book, and was also intrigued by the precise experiments in Manchester of James Prescott Joule, which attempted to demonstrate the conversion of mechanical work into an equivalent amount of heat, by measuring tiny rises in the temperature of a water bath agitated by paddles driven by falling weights. Thomson heard a presentation from Joule at an Oxford meeting of the British Association in 1847. Joule explained his experiments of two years before, when he had measured the work required to raise the temperature of one pound weight of water by one degree Fahrenheit as 817 'foot-pounds' and also deduced, by backwards extrapolating measurements of the temperatures of expanding gases, that there should be a 'zero' of pressure and hence temperature at 480°F below the freezing point of water. Thomson was *"tremendously struck with"* Joule's paper. They became life-long friends. Within a year Thomson had realised the need for defining a 'thermometric' scale of temperature, which would have two advantages. First, it would avoid the dependence on the properties of particular substances used in thermometers, such as mercury and air, which had coefficients of expansion which were temperature dependent, and therefore not ideal. Secondly and more fundamentally, by basing the scale on the 'absolute zero of temperature' rather than arbitrarily basing it on the freezing point of water, the temperature of a substance became the defining measure of the amount of heat it contained, and the difference in the *absolute temperatures* of two substances defined the maximum amount of heat that could flow between them. Using the centigrade temperature scale of Celsius proposed a century before, Thomson identified absolute zero to be close to -273°C. The temperature -273.16°C is now known as 0K, zero degrees Kelvin. Within a year Thomson published these ideas in the *Proceedings of the Cambridge Philosophical Society*, titled *'On an Absolute Thermometric Scale founded on Carnot's Theory of the Motive Power of Heat and Calculated from Regnault's Observations'*.

Thomson was now focused on the subject that he named 'thermodynamics' and read a paper to the Royal Society of Edinburgh on January 2nd 1849, called *'An account of Carnot's theory of the motive power of heat; with numerical results deduced from Regnault's experiments on steam'*, in which he summarised Carnot's work, still accepting at this point Carnot's axiom that caloric is indestructible, and that the work done by heat engines was due to the caloric 'falling' through the engine from hot to cold without loss, analogous to water falling through a water-wheel. Yet at the same time Thomson highlighted the experiments of Joule which showed that heat is generated by the friction of fluids, and by electric current, strongly suggesting the equivalence and the direct convertibility of other forms of energy to heat, and potentially *vice versa*. Thomson was in the process of being convinced by Joule's conclusions, and in a footnote wrote the crucial words: *"When 'thermal agency' is thus spent in conducting heat through a solid, what becomes of the mechanical effect which it might produce? Nothing can be lost in the operations of nature - no energy can be destroyed. What effect is then produced in place of the mechanical effect which is lost? ...no answer can be given in the present state of science...It might appear that the difficulty could be avoided, by abandoning Carnot's fundamental axiom, a view which is strongly urged by Mr. Joule. If we do, however, we meet with innumerable other difficulties".* That passage contains the first use of the word *'energy'* in its modern context, and the paper also introduced the term *'thermodynamic* engine', heralding the birth of a new science.

A few days after the Edinburgh presentation, huge domestic sadness struck. An epidemic of cholera had engulfed Glasgow, and William's beloved father had been infected. On 12th January William wrote *"We have lost our father. He died this afternoon...on the Sunday night he became delirious, and since that time he has been gradually sinking in strength. I could not believe last night at this time that we were to lose him...it is a terrible and irreparable loss, and a sad void is now left."* The guiding light of his son's early life was 62 years old. After this horrible, sudden loss, Thomson continued to live at No. 2, The College. After a pause, he threw himself again into the study of heat, now joined in this interest by his brother James, who had graduated with great distinction from Glasgow University and was establishing an impressive career in engineering.

Meanwhile, Joule had continued to refine his experiments and, from measurements on the condensation of air, had concluded that the heat of gases consisted purely of the *vis viva* (active force) of their particles. The prodigious mathematician and ingenious Glasgow engineer William John Macquorn Rankine had developed, but not published, a similar idea in his *'hypothesis of molecular vortices'* which proposed that it was the centrifugal forces of whirling particles which endowed gases with their elasticity and content of heat. Thomson now accepted Joule's findings and read another paper to the Edinburgh Society called *'How must the theory of the heat-engine be modified, supposing that heat is not a substance, but a mode of motion?'* Rankine gave the answer with a brilliant mathematical exposition of his vortex theory in a paper *'On the mechanical action of heat, especially in gases and vapours'*, read to the Royal Society of Edinburgh in February 1850. It provided the theoretical underpinning for Joule's experimental results, and in an almost casual aside made clear Rankine's key realisation regarding heat-engines: *"Carnot, in fact considers heat ...is incapable of increase or diminution. According to these principles, a body, having received a certain quantity of heat, is capable of giving out not only all the heat it has received, but also a quantity of mechanical power which did not before exist. According to the theory of this Essay...which regards heat as a modification of motion, no mechanical power can be given out in the shape of expansion unless the quantity of heat emitted is* <u>less</u> *than the quantity of heat originally received: the excess... to appear as expansive power, so that the sum of the 'vis viva' in those two forms continues unchanged."* Rankine had arrived at the conclusion that it is not heat that is conserved in a heat-engine, but the *vis viva* in the system - what we would now recognise as 'energy'. At almost exactly the same time another scientist was drawing similar conclusions, and would become known as the third father of thermodynamics, alongside Thomson and Rankine. The German physicist Rudolf Clausius had absorbed the lessons of Joule's experiments and the hints in Thomson's footnote, and accepted as given the equivalence of heat and work, and that heat was a measure of the internal motions of particles in a body. In his paper *'On the Moving Force of Heat and the Laws regarding the Nature of Heat itself which are deducible therefrom'* (communicated to the Berlin Academy of Sciences in February 1850 and eventually published in English in July 1851), he considered Thomson's *"innumerable difficulties"* to be not so great as Thomson imagined, and began his first thoughts on what would later be termed 'entropy' - wasted, dispersed heat produced in any irreversible process. He showed how discarding Carnot's axiom of indestructible caloric was quite compatible with the rest of Carnot's theory if heat and work are considered interchangeable.

It was time for Thomson to draw all this together, which he did in his memoir *'On the Dynamical Theory of Heat...'* read to the Royal Society of Edinburgh in March of 1851. He traced the idea of heat-as-motion, from Sir Humphrey Davy, through the experiments of Joule and Julius Robert von Mayer, to the theory of Carnot and Clapeyron and its correction by Rankine and Clausius. Developing the mathematics throughout the following year, he definitively adopted the term *'energy'* and codified what we now call the First and Second Laws of Thermodynamics, best expressed simply as:

1st Law: the total energy of an isolated system is constant, and energy can be neither created nor destroyed. For a closed system, any increase in internal energy is equal to the heat supplied minus the work done by the system on its surroundings.

2nd Law: it is impossible for a self-acting machine, unaided by any external agency, to convey heat from one body to another at a higher temperature.

He arrived at the expression (as did Rankine) that defines the maximum possible efficiency of a heat engine in terms of the absolute temperatures of the heat source (T'), pronounced T *dash*, and the cold 'sink' (T):

Ratio of maximum work produced to input heat =

$$\frac{(T' - T)}{T'}$$

which was immortalised in the rhyme of one of his students, James Napier:

When you yourself once taught me

Heat's greatest work to know

Wasn't it T dash minus T

With T dash down below?

Thomson went on the next year to discuss what he called *"the universal tendency of nature to the dissipation of mechanical energy"*, a description of the concept of universally increasing entropy. With hindsight, it seems obvious that Carnot's 'caloric axiom' was wrong, and strange that it took Thomson so long to accept that. To be fair, in the early 1850s, Thomson was busy working also on other subjects, and perhaps distracted too. Aged 28, already famous, and settled in his Glasgow Chair, Thomson had renewed his acquaintance with Margaret Crum, a beautiful and accomplished girl of 22 he had known since boyhood. In the summer of 1852 he wrote to his sister Elizabeth, who knew Margaret well. *"I have a piece of news to tell you which I think will please you as much as it will surprise you. Margaret Crum has consented to be mine, and from today we are engaged to be married. I came up from Largs this afternoon, and rode out to Thornliebank to tell Mr. and Mrs Crum and ask their consent, which was given with the greatest goodwill."* A week after the engagement Margaret wrote to Elizabeth: *"...as I told Mrs. Gall, I feel that in William's love for his sisters and her, lies my best security for the continuation to me of those feelings on which the happiness of my life must now depend."* After the wedding and a happy honeymoon in Wales, followed by a holiday in the Mediterranean the next year, Margaret's health began sadly to decline, and she required much care from William for many years. In all of 1853 Thomson published just three notes in the *Cambridge and Dublin Mathematical Journal* and the *Philosophical Magazine*, and there are no further entries in his mathematical notebook until 1855.

Illustration 48: William Thomson, the future Lord Kelvin

At Glasgow, Thomson went on to establish a teaching and research laboratory in his department, the first of its kind in any British university, helped in its construction by his favourite students including John Kerr. His subsequent achievements in various fields - electricity, magnetism, the behaviour of gases, geology, telegraphy, and instrumentation - established him as one of the greatest physicists of all time. He engineered the first submarine telegraph cable across the Atlantic, and was on board the *Great Eastern* as consultant when the huge ship was used to lay the first trans-atlantic cable in 1866. For this and all his other pioneering scientific contributions, he was knighted by

Queen Victoria. Sadly though, Margaret's health had been in almost continuous long-term decline and she died in June 1870 at their house in Largs, specially rented to aid her hoped-for recuperation.

Sir William had made many innovations in the field of telegraph systems and instrumentation, and was reaping the financial rewards for his work and patents. He donated a share of his gains to Glasgow University "for promoting the cultivation of experimental science". In 1870 the University underwent a huge transformation, relocating from its old outdated College buildings near the High Street, some dating from 1451, to a leafy site in the west of the city at Gilmorehill, overlooking the River Kelvin and Kelvingrove. His wife's death prevented Thomson from taking any active part in the move of his laboratory to the new buildings, but he was soon installed, and resumed work in teaching physics with a new interest in geological research. He rejected Hutton's and Lyell's ideas of an indefinitely old Earth, and based on measurements of underground temperatures, estimated the age of the Earth at approximately 100 million years - far too young compared even to the under-estimates of Darwin and the contemporaneous work of Thomas Huxley.

To William's great satisfaction, his brother James was appointed to the chair of engineering at Glasgow in 1873, after William John Macquorn Rankine's sad and untimely death the previous year. Around this time Thomson became a keen yachtsman, and fell in love for a second time, with Frances Blandy, daughter of a landowner in Madeira. They were married in Funchal in 1874. After many years of marriage and yet more distinctive scientific contributions to the standardisation of scientific units of measurement, and to electric light and dynamos, Thomson's house was one of the first to be lighted by electricity. He was made a peer on New Year's Day 1892 and took the name Baron Kelvin of Largs. The year was marked with huge sadness too, with the deaths of William's dear brother James, at the age of 70, together with James's wife and one of their daughters, all in the same week.

Lord Kelvin retired from his Glasgow professorship on his 75th birthday in 1899, but remained active in business and the House of Lords, while enjoying 'whimsical experiments' and international travel. He corresponded widely, particularly with his great friend Hermann Helmholtz, worked closely with Lord Rayleigh, and met Edison, Westinghouse, Tesla, Marconi, and Pierre and Madame Curie, whose work caused him to reassess the cherished principle of the conservation of energy. He became Chancellor of the University of Glasgow in 1904. Lord Kelvin died in November 1907 after an astoundingly prolific scientific career spanning most of the 19th century. He was buried with great honour in Westminster Abbey, beside the graves of Sir Isaac Newton and Charles Darwin.

40 The beautiful equations that changed the world: James Clerk Maxwell (1831 –1879)

The brilliant experimental work of Michael Faraday in the 1820s and 1830s established the basic facts about the interactions of the mysterious forces of magnetism and electricity. Together with the work of Ørsted, Ampere, Coulomb and Volta, Faraday's experiments had demonstrated four fundamental facts. Number one: electric charges attract

Illustration 49: James Clerk Maxwell - engraving by G.J. Stodart

or repel each other with a force inversely proportional to the square of the distance between them. Secondly, magnetic poles always come in pairs and attract or repel with the same dependence on their distance of separation. Thirdly, an electric current in a wire creates a circular magnetic field wrapped around it, with its direction of 'rotation' dependent on the direction of the current. And lastly, a magnet moving through a loop of wire induces an electric current in the wire, with the direction of current flow depending on the direction of the magnet's movement. These four discoveries enabled the invention of a range of ingenious electrical machines, including early electric motors and generators (or dynamos). But what was really going on? And how could these distant interactions of electricity and magnetism be understood and explained? Faraday was a superb experimentalist, but he did not have the mathematical ability to create a unified theory of electricity and magnetism.

The man who did, and who made some of the most significant advances in physics of all time, was James Clerk Maxwell. He thought deeply about many aspects of the physical world, and his acute insights ranged across the subjects of optics, colour perception, gas kinetics, and thermodynamics. But it was his beautiful and elegant formulation of a complete theory of electromagnetism that changed physics forever, and enabled a new era of radical thought that would be exemplified by Albert Einstein's Special Theory of Relativity.

Maxwell was born in Edinburgh, the second child of John Clerk Maxwell, a prosperous lawyer and landowner, and his wife Frances Cay. His family had distinguished forebears on both his father's and mother's sides. His father was the younger brother of Sir George Clerk of Penicuik, who held government roles in the administrations of Lord Liverpool, the Duke of Wellington and Sir Robert Peel. John's great-grandfather, Sir John Clerk, had been a talented composer and scientist, a fellow of the Royal Society, a Baron of the Exchequer of Scotland and a Commissioner of the Union. Frances was a daughter of Robert Hodshon Cay, principal judge in the High Court of the Admiralty in Scotland. In a complicated story of inheritance, James' father became heir to a Maxwell-owned estate called Glenlair, in rural Galloway, and added the Maxwell name to his own. John Clerk Maxwell and Frances were in their mid-thirties when they married. Their first child, Elizabeth, died in infancy, and when James was born his parents doted on him. His early upbringing at the Glenlair Estate must have been idyllic. James' curiosity in the workings of all things was striking from a young age, forever demanding *"show me how it doos"* and asking, *"what's the go o' that?"* Frances gave him all his early education, but the family's second tragedy was imminent when she was diagnosed with abdominal cancer. Soon after an attempted remedial operation, Frances died, aged 47. James was just eight years old. After an unhappy couple of years of home schooling from a young private tutor, James was sent to Edinburgh to live with his Aunt Isabella, and was enrolled at the Edinburgh Academy, one of the best schools in Scotland. He joined a class of boys who had been at the school for a year, and with his rural Galloway accent and home-designed country clothes he was soon a figure of fun, and nicknamed 'Dafty'. His aunt soon sorted out his attire and with frequent visits from his father, James was soon settled and excelling at school in all subjects. He made two special life-long friends: Lewis Campbell, who became professor of Greek at St Andrews and would be a biographer of Maxwell; and Peter Guthrie Tait, who became one of the world's top physicists in his own right, and would be professor of natural philosophy at Edinburgh University over a span of four decades, succeeding the eminent James Forbes.

Forbes, indeed, was the no-doubt surprised recipient of Maxwell's first mathematical paper, written when James was just fourteen. The subject was a mathematical treatment of the many geometrical shapes - ellipses and irregular ovals - that can be drawn using pins and string, and it expanded on discoveries that had been made by René Descartes. James' father submitted the paper to Forbes who replied, *"I have looked over your son's paper carefully, and I think it is very ingenious, certainly very remarkable for his years; and I believe substantially new."* Forbes himself read the paper to the Royal Society of Edinburgh since James was so young. As a result, when Maxwell enrolled two years later at the University of Edinburgh, in the autumn of 1847, Forbes managed to persuade James' father that his son's future lay in mathematics and physics, and not the study of law.

Earlier that year Maxwell visited the laboratory of William Nicol, the inventor of the eponymous polarising prism, which ignited James' interest in optics and the theory of colour vision. He pursued the interest during his three years of study at Edinburgh, and in his final year published impressive papers on 'rolling curves' and on the optical effects of strain in elastic materials, demonstrating that he was already, aged 18, an accomplished physicist and mathematician. He moved on to St. Peter's College in the University of Cambridge, where his friend P.G.Tait was an undergraduate, and then, on the recommendation of Forbes, he moved to Trinity College, to study with William Hopkins, then the top mathematics 'coach' for the Cambridge mathematics Tripos exams. In 1854, Maxwell graduated as Second Wrangler and was joint winner of the prestigious Smith prize - a very similar outcome to that of William Thomson, nine years before, but not quite as good the performance of Senior Wrangler, one P.G.Tait, two years before - no doubt much to Tait's satisfaction and amusement. The examination room was not the best forum for Maxwell's formidable inventiveness and originality. He was highly popular at Trinity and his jokey humour, kindness and helpfulness to friends and students were widely appreciated. After graduation, he remained at Cambridge as a tutor and scholar - and was elected a fellow of Trinity College in 1855. During these years, Maxwell thought about the compatibility of his strong, Presbyterian religious faith with his philosophy of science, and supported the Christian Socialist group of Frederick Denison Maurice. He was a committed supporter of Richard Buckley Litchfield who ran the Working Men's College in London, and Maxwell also taught classes for working men in Cambridge.

All this still left plenty of time for research, and Maxwell resumed his interest in colour vision and colourimetry. He invented one of the world's first ophthalmoscopes to study the retinas of animals and humans, trying to answer the question he had asked as a three-year-old: *"but how d'ye know it's blue?"* He had the crucial realisation that mixing coloured pigments gives entirely different results from mixing colours of light. For example, as artists and children know well, mixing blue and yellow paint gives green. However, mixing blue light with yellow light gives pink. The explanation is that pigments selectively *absorb* light colours; a blue pigment looks blue because all other colours except blue are absorbed and subtracted from the incident light. Conversely, mixing coloured light is an additive process. Through a series of ingenious experiments using coloured spinning tops and discs, following methods he learned from Forbes, Maxwell concluded that any perceived colour can be created by combining the 'primary' colours of red, blue and green light. He devised a 'colour triangle' which showed which proportions of red (R), green (G) and blue (B) were needed to make any desired colour - and the RGB primary colour system became the basis for modern television and display screens. Maxwell would go on a few years later, at a demonstration in the Royal Institution, to project the world's first colour photograph - a picture of a tartan ribbon, produced by superimposing three images, taken in turn through red, green and blue filters.

The train of thought which led to Maxwell's most towering achievement began in 1855, when he considered the state of the known laws of electricity and magnetism. In a paper read to the Cambridge Philosophical Society called *'On Faraday's Lines of Force'* he introduced his subject by reviewing the various disjointed electrical laws and patchy deductions from magnetic experiments, saying *"The present state of electrical science seems peculiarly unfavourable to speculation...the first process therefore in the effectual study of the science, must be one of simplification and reduction of the results of previous investigation to a form in which the mind can grasp them. The results of this simplification may take the form of a purely mathematical formula or of a physical hypothesis."* Faraday had concluded from his many experiments that magnets and electrical charges interacted with each other through 'lines of force' which extend throughout the spaces between them. The opposing view was that electric charges and magnets interacted by 'action at a distance' with nothing interposed in the spaces between. Faraday was ridiculed by some of the advocates of this view including Sir George Airy, the Astronomer Royal. Maxwell was on Faraday's side, but saw the need to explain the behaviour of the 'lines of force' in detail, mathematically, simply, and in a way that made sense of the piecemeal laws of electricity and magnetism. He started by drawing an analogy with the flow of fluids - and imagined a weightless, incompressible fluid flowing through a porous medium. The streamlines of the flow determined the direction of Faraday's lines of force, the porosity represented the electric and magnetic properties of the intervening material, and the speed of the flow was related to the density or 'flux' of the force-lines. The force was strong where the flux was dense, weak where the flux was diffuse, and the flow was caused by a difference in 'pressure'. These analogies would nudge his thoughts towards his greatest achievement - a complete mathematical description of electromagnetism.

In February 1856 Maxwell received a letter from his Edinburgh mentor, Professor Forbes, alerting him that the chair of natural philosophy had become vacant at Marischal College, University of Aberdeen. Forbes encouraged Maxwell to apply, stressing that he had no role in making the appointment, which was in the gift of the Home Secretary and the Lord Advocate. Though he would be sorry to leave Cambridge, Maxwell decided to apply, and the shorter academic year in Scotland would allow him to spend more time with his dear father, whose health was in decline. His rivals for the post included his old friend P.G. Tait, who was by then professor of mathematics at Queen's College, Belfast, but who wanted to return to Scotland and continue his work in physics. While awaiting the decision, Maxwell spent the Easter vacation at Glenlair, nursing his father, who sadly and suddenly died one morning after a troubled night. Amidst his grief, Maxwell assumed his role as Laird and supervised the funeral and the general business of the estate and its workers. When he returned to Cambridge he heard that he had won the competition for the professorship at Aberdeen.

Marischal College was large, grey, granite and austere. Its professors shared most of those characteristics. But the new young professor - aged only 25 - was given a friendly welcome, though he wrote to Lewis Campbell, *"No jokes of any kind are understood here. I have not made one for two months, and if I feel one coming on I shall bite my tongue."* In his inaugural professorial lecture, Maxwell avoided his usual humour and stressed the importance of the empirical approach to science. *"I have no reason to believe that the human intellect is able to weave a system of physics out of its own resources without experimental labour. Whenever the attempt has been made it has resulted in an unnatural and self-contradictory mass of rubbish."*

Maxwell made three major advances in his life and work while in Aberdeen. First, he fell in love with Katherine Mary Dewar, the daughter of the College Principal, and the love was mutual. They were engaged in February 1858 and married that summer, with Lewis Campbell as best man, as Maxwell had been his a few weeks before. At their engagement, Maxwell wrote to his Aunt Jane: *"This comes to tell you that I am going to have a wife. I am not going to write out a catalogue of qualities, as I am not fit; but I can tell you that we are quite necessary to one another, and understand each other better than most couples I have seen. Don't be afraid; she is not mathematical; but there are other things besides that, and she certainly won't stop the mathematics. The only one that can speak as an eyewitness is Johnnie, and he only saw her when we were both trying to act the indifferent. We have been trying it since, but it would not do, and it was not good for either."* Katherine was seven years older than James, and their marriage would be close and happy, though childless.

Secondly, at Aberdeen, Maxwell became intrigued by the problem of explaining the stability of the rings of Saturn, which had been set by the University of Cambridge as the target for its so-called Adams Prize in 1855. The question was whether it could be shown that the rings would be stable if they were fluid, or solid, or composed of separate bodies. The problem had been attacked by many mathematical astronomers without success, including the author of *La Mécanique Céleste*, Pierre-Simon Laplace. Over a period of two years Maxwell used his mathematical ingenuity to show that symmetrical solid rings would be unstable; that fluid rings would break up into separate 'blobs'; and that the uniform symmetrical rings that we observe could only be explained if they consisted of separate particles or bodies orbiting independently. For this he was awarded the Adams Prize in the year of his marriage, with the Astronomer Royal, Sir George Airy, describing Maxwell's analysis as *"One of the most remarkable applications of mathematics to physics that I have ever seen."* NASA's *Voyager* fly-past missions of Saturn in the 1980s revealed the rings to have the structures that Maxwell's mathematics predicted.

The third breakthrough made in Aberdeen was Maxwell's successful prediction of the distribution of molecular velocities in gases. The 'kinetic theory of gases' postulated that gas consisted of molecules moving around at enormous speed, a speed which increased with the temperature of the gas. The molecules collide with the walls of their container, creating the pressure of the gas. This kinetic model was consistent with the gas laws derived from experiments and formalised by Robert Boyle (the inverse dependence of pressure on volume), Jacques Charles (that the volume of gas at constant pressure is proportional to its absolute temperature), and Joseph Gay-Lussac (at constant volume, the gas pressure is directly proportional to its absolute temperature). So far so good. But there were problems over the details of the molecular movements. To explain the observed pressures at normal temperatures, the molecules would need to be travelling at several hundred metres per second. Yet when gases diffuse or intermingle, they do so slowly - for example, as the scent from a bottle of perfume spreads gradually when the bottle is opened. Rudolf Clausius proposed the solution. The molecules must be colliding with each other with huge frequency, travelling between collisions only a short average distance he called the 'mean free path'. Clausius supposed that the molecules could be moving at different speeds at different moments, but for simplicity assumed that at a given temperature all molecules of the same kind travelled at the same speed. Maxwell was intrigued, and set himself the task of deriving a theory of molecular speeds that would fit the facts.

He began by intuiting that a molecule moving in three dimensions with velocities x, y and z could have different velocities in each direction, and would have an overall speed 'v' given by

$$v^2 = x^2 + y^2 + z^2$$

Also, he saw that the number of molecules with a particular velocity in one direction (say x) will depend on the number of molecules that have a total speed v, but will not depend on the numbers having particular velocities in the y and z directions. Furthermore, the general shape of the velocity distributions in the x, y and z directions must be the same, since there is nothing special to distinguish one direction from another. The mathematical conclusion was clear: the distribution of velocities in each x, y and z direction was a bell-shaped curve centred on zero velocity, (since the molecules can move equally in a 'positive' or 'negative' direction). This was the well-known 'normal' distribution - but one that flattened and broadened as the gas temperature increased. The resulting distribution of overall molecular speed had a more interesting shape, shown in Illustration 50, for a calculation assuming one million oxygen atoms at three different temperatures of -100°C, 20°C and 600°C. The curves show the number of molecules having any particular speed (in metres per second), and the area under each curve represents the one million total number of molecules. This is now known as the Maxwell-Boltzmann distribution, in honour of both Maxwell's derivation and of the contributions later of Ludwig Boltzmann who together with Maxwell founded a whole new sub-discipline of thermodynamics - statistical mechanics. Maxwell's theory correctly predicted one surprising feature of gases that was confirmed by experiment: that the viscosity of a gas - the frictional drag it imposes

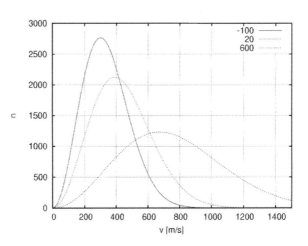

Illustration 50: Maxwell-Boltzmann distribution of speeds calculated for 1 million oxygen molecules at 3 temperatures

on a body moving through it - is independent of the gas pressure, a counter-intuitive result that surprised even Maxwell. Yet it was many decades before the kinetic theory of gases was fully accepted, because many physicists, especially on the European continent, still denied the physical reality of atoms and molecules. In Great Britain, most accepted the existence of molecules, but ascribed to the Newtonian view that gas pressure resulted from the static repulsion between molecules, not from the high-speed collisions and random walks of the molecules in the kinetic theory.

Some of the rude realities of life intruded on Maxwell in the year 1860. It was a strange fact that of Scotland's five ancient universities, two were in Aberdeen. 'Marischal College, University of Aberdeen' was founded in 1593 by George Keith, 5th Earl Marischal of Scotland, without Papal consent and modelled on the Protestant arts colleges and universities of northern Europe. 'The University and King's College of Aberdeen', established almost a hundred years earlier by Papal bull and royal decree of King James IV, modelled itself on the University of Paris, the *alma mater* of its founder, Bishop Elphinstone. Not surprisingly, a merger of the two rival universities had been mooted, not least on the grounds of economy, and a Royal Commission decided on a fusion that would halve the numbers of professors in a new University of Aberdeen. Maxwell's rival for the united chair of natural philosophy was the well-established King's College Sub-Principal, David Thompson. On the grounds of 'last in, first out', and given that Maxwell's short tenure had not yet qualified him for a pension or pay-off, he was made redundant. Aberdeen had lost one of the greatest physicists of all time.

Almost simultaneously, James Forbes retired his chair at Edinburgh to become Principal of St Andrews University. Maxwell applied for the vacancy and so did P.G.Tait, who on this occasion won the contest and returned to Scotland. Out of a job, Maxwell completed his work on gas kinetics for publication in the *Philosophical Magazine* as '*Illustrations of the Dynamical Theory of Gases*' and was soon aware of the vacant chair in natural philosophy and astronomy at King's College, London. He applied, and won the position beginning in the academic year 1860-61. Even before he could move to London, another problem struck. Always an enthusiastic rider, Maxwell attended a summer market fair to buy a horse for his new wife Katherine. After returning home with a beautiful bay pony, he fell violently sick and developed a raging fever. He had contracted smallpox. Maxwell was seriously ill, and he was forever sure that without dedicated nursing from Katherine, he would have died. But by October he was on the mend, and he, Katherine and Charlie (the pony) moved to Kensington Gardens near Hyde Park in London - perfect for riding out in the afternoons.

The year had been difficult, even traumatic, but not without great successes. Maxwell was awarded the Rumford Medal, the highest honour of the Royal Society, for his work on colour science, and he settled quickly at King's College. His teaching load was a bit lighter than at Aberdeen, though the academic year was longer, and he continued his customary evening lectures for 'artisans' once a week. Maxwell seems to have been well-loved by all his students, but his teaching style was not to everyone's taste. He was always well prepared, but despite his mathematical genius, he sometimes made algebraic mistakes on the blackboard (as he did also in some of his published papers) and was fond of making exuberant analogies which often did more to mystify than to clarify. In his research, he had returned to think about electricity and magnetism, where a unifying theory was still sorely needed, and one analogy in his mind was proving particularly fruitful - at least to himself.

Maxwell needed a way to imagine how lines of magnetic force could exert an attractive tension along their length but a repulsive outward pressure at right angles to the lines of force. He imagined space to be filled with tiny, close-packed spherical cells, which he called 'vortices', in contact with each other and free to spin. When spinning, they would tend to expand around their 'equators', exerting sideways pressure, and contract along the axis of spin, creating a tension. If the spin axes were aligned with the lines of force, this mental model provided the right behaviour for magnetic lines of force. Now, Maxwell imagined the cells not to be in direct contact but separated by smaller particles which acted like tiny ball-bearings so that when one cell rotated, only some of its neighbours would be counter-rotating. He imagined the small particles to be particles of electric charge, and it was the current of particles rolling through the mesh of cells that set them rotating or counter-rotating - creating the magnetic lines of force in either direction. Whether you find the analogy mystifying or clarifying, for Maxwell it was just the thought experiment he needed to see how the linear electric current in a wire could create the rotatory magnetic forces around it. He wrote up the analogy in 1861 with full mathematical rigour as *'On physical lines of force'* published initially in two parts in the *Philosophical Magazine*. Now, as he might have joked himself, he was on a roll.

The next months were spent thinking about the implications of his vortex analogy, leading to one the most important scientific breakthroughs of all time. Maxwell never regarded the spinning cells as a physical reality, but as a mathematical model they prompted some fundamental realisations. Space itself could form the link between linear electric currents and rotational magnetic forces. Further, electric currents could briefly flow in empty space and in insulators, wherever there was change in the gradient of the electric field at that point. Maxwell called this *'displacement current'*. And lastly, the whole web of spinning cells and electric particles would sustain waves that would ripple transversely across the network in response to a disturbance at any one point. Light was known to be a transverse wave. Could it possibly be that light was a wave of electromagnetism? Maxwell did some calculations. If he hypothesised that his imaginary spinning cells were perfect solids, waves would propagate through the network at 310,740 kilometres per second. The best measured value of the speed of light, by Hippolyte Fizeau, using an intense light source and a mirror 8km distant, was 314,858 kilometres per second. Astounding. In the third part of *'On physical lines of force'* Maxwell wrote calmly *"The velocity of transverse undulations in our hypothetical medium...agrees so exactly with the velocity of light calculated from the experiments of M. Fizeau, that we can scarcely avoid the inference that light consists in the transverse undulations of the same medium which is the cause of electric and magnetic phenomena."*

The hypothetical vortex-cell analogy had served its purpose and it was not a model that could be sustained. Maxwell's friend, the mathematician Cecil James Monro, observed, *"... a brilliant result. But I must say I think a few such results are needed before you can get people to think that every time an electric current is produced a little file of particles is squeezed along between two rows of wheels."* Quite. To take the theory further, Maxwell started all over again with entirely new mathematics. In a superb synthesis of all he had discovered, he defined the two critical quantities: **E** and **H**, respectively the electric and magnetic field strengths at a point in space, which were vector quantities (that is, having direction as well as magnitude). Then, the rate of change of these quantities with time t: $\partial E/\partial t$ and $\partial H/\partial t$. Lastly a constant c, the ratio of electromagnetic to electrostatic charges. To explain the relationships between the linear and rotational forces, he defined two entirely new mathematical functions: the 'divergence' (div for short) to measure the tendency of a force to act and spread outwards (div positive) or focus inwards (div negative). And 'curl' to describe the tendency of a force to loop or curl around a given point - with positive or negative curl defining the direction of looping. The result was a complete description of all electromagnetic phenomena that can be summarised in the four Maxwell Equations, whose beauty, simplicity and elegance can be appreciated even by the least mathematical. For a point in empty space with no charges present, they are:

$$\text{div } \mathbf{E} = 0 \qquad (1)$$

$$\text{div } \mathbf{H} = 0 \qquad (2)$$

$$\text{curl } \mathbf{E} = -(1/c)\, \partial \mathbf{H}/\partial t \qquad (3)$$

$$\text{curl } \mathbf{H} = (1/c)\, \partial \mathbf{E}/\partial t \qquad (4)$$

The first two equations say that in this case no electric charges or magnetic poles are present, and imply the familiar inverse square law of field strength with distance, as stated by the laws of Coulomb and Gauss. Equation (3) relates a changing magnetic field to the circular electric field it induces, and is equivalent to Faraday's Law of electromagnetic induction in a wire loop. Equation (4) relates a change in electric field or current to the circular magnetic field around it, and reduces to Ampere's Law. However, and amazingly, if equations (3) and (4) are taken together they describe the simultaneous, self-reinforcing propagation of magnetic and electric fields at a speed c, which turns out indeed to be the same as the measured speed of light. Maxwell's *tour de force* was published in seven parts as *'A dynamical theory of the Electromagnetic field'* starting with a presentation to the Royal Society in December 1864. Most of his contemporaries were astonished and bemused. Faraday, never a mathematician, had long pleaded with Maxwell to describe his results in plain English. Even William Thomson struggled to grasp the implications of the new mathematics. The theory was far ahead of its time. It would be more than 20 years before Heinrich Hertz demonstrated the reality of electromagnetic waves, and opened the door to a new century of technological innovation. With hindsight, the synthesis of electricity and magnetism in Maxwell's beautiful equations was the supreme achievement of 19th century physics. His equations prompted new measurements of the speed of light and began the fruitless search for the 'luminiferous ether' that should sustain the electromagnetic waves, leading to the realisation that the speed of light was a universal constant. Einstein kept a photograph of Maxwell in his study at Princeton and when he was asked if, in developing the Special Theory of Relativity, he *"stood on the shoulders of Newton"* he replied *"No, on the shoulders of Maxwell."*

Maxwell's pursuit of the complete theory of electromagnetism had taken nine years, five of them while in London, and there was much more to do. In 1865 he resigned his professorship at King's College, and handed over the chair to William Grylls Adams, younger brother of

John Couch Adams of the eponymous Prize won by Maxwell seven years previously. While in London, he made yet another immense contribution to science. The whole system of electric and magnetic units of measurement had grown up through various experiments and different definitions had been made across the world. It was a mess. The British Association for the Advancement of Science had asked Maxwell to lead a team to sort it out. In fact, the problem went far beyond electricity and magnetism. General terms like 'work', 'force' and 'power' had no commonly agreed definitions either. Maxwell picked as his team two other Scots, Balfour Stewart, director of the Kew Observatory, and Fleeming Jenkin, later to be professor of engineering at UCL and Edinburgh. The team devised the first consistent system of units to be internationally accepted, based on centimetres, grams and seconds - the 'cgs' system. Maxwell went further, showing that all physical quantities could be expressed as combinations of mass (M), length (L) and time (T). For example, any unit of velocity must have the dimensions L/T, acceleration must be L/T^2 and any unit of force (which equals mass x acceleration by Newton's 2nd Law of Motion) must have the dimensions ML/T^2. The technique of 'dimensional analysis' is now second nature to engineers and scientists in checking the self-consistency of their equations, and is commonly used in finance and accounting too.

In the spring of 1865, James and Katherine Clerk Maxwell left London and returned home to Glenlair. This was certainly not retirement. Maxwell was not yet 34 years old. James and Katherine enjoyed life and relaxed by riding in the afternoons and travelling abroad. But James worked hard on improving the Glenlair estate and on his major books, *Treatise on Electricity and Magnetism* and *The Theory of Heat*, as well as publishing many more research papers brimming with originality and insight - in optics, colour science, the kinetic theory of gases, topology, and the diagrammatic analysis of static forces (this last building on the work of Rankine). Maxwell's thoughts on the operation of mechanical governors in regulating the speed of steam engines led him to the concept of negative feedback, and he laid down the theory used in modern control systems. He was in frequent scientific correspondence, exchanging ideas and jokes with William Thomson at Glasgow and P.G. Tait at Edinburgh, and in more formal communication with Helmholtz at Heidelberg and Boltzmann in Austria. He attended Cambridge annually where the University had asked him to be the examiner of its Mathematical Tripos exams.

Cambridge lured him back full-time in 1871, with the offer of perhaps the only post that would have coaxed him away from Glenlair - its professorship of experimental physics, with the task of building and directing a brand-new teaching and research laboratory, to be called the Cavendish. The University had neglected its laboratory facilities, and needed to catch up with the competition. It had approached William Thomson who had established a world-class laboratory at Glasgow, but he declined. So did Helmholtz. Maxwell was third choice. He designed the laboratory and acquired its instruments, paid for largely by its patron, the Duke of Devonshire (great-nephew of Henry Cavendish), and by Maxwell himself. Learning from visits to the best university laboratories, including Thomson's, Maxwell specified a building full of light, with tall windows and large spaces to accommodate equipment requiring great height or length. In his inaugural professorial lecture, he attacked the view that physics was largely 'done', which even Thomson succumbed to in his last years as Lord Kelvin. Maxwell said *"...the opinion seems to have got about that in a few years all the great physical constants will have been approximately estimated, and that the only occupation which will be left to men of science will be to carry on these measurements to another place of decimals...but the history of science shows that...she is preparing materials for the subjugation of new regions..."* The Cavendish Laboratory was opened in 1874, and 23 years later it would be the site of J.J. Thomson's discovery of the electron.

Enjoying his return to Cambridge, Maxwell was his irrepressible self, recruiting researchers; encouraging students; writing whimsical poetry that poked fun at colleagues; and welcoming and advising visitors from abroad. He joined a convivial essay club of dons and professors, and as always dispensed thoughts far ahead of their time. Anticipating the future science of Chaos Theory and the 'Butterfly Effect', he dismissed the view that scientific laws imply a deterministic universe in which the future is predictable with unlimited precision, writing *"...when an infinitely small variation...may bring about a finite difference in the state of a system, the condition is unstable. It is manifest that the existence of unstable conditions renders impossible the prediction of future events, if our knowledge of the present state is only approximate..."*

In the spring of 1877 Maxwell began to suffer heartburn and stomach pains. He returned to Glenlair, where his Edinburgh doctor diagnosed abdominal cancer, the same disease that had killed Maxwell's mother at 47 years of age. He was advised to return to Cambridge for treatment and pain relief, which he did, but died there, attended by Katherine, in early November. He was 48 years old. As not only a brilliant genius, but as a kind, humorous and much-loved man, the grief of his friends was great, and their epitaphs were heartfelt. On his scientific legacy, the last words are perhaps best left to two of the greatest physicists of the 20th century. Richard Feynman said, *"From a long view of the history of mankind, seen from, say, ten thousand years from now, there can be little doubt that the most significant event of the 19th century will be judged as Maxwell's discovery of the laws of electrodynamics."*

It was summed up by an observation of Albert Einstein. *"One scientific epoch ended and another began with James Clerk Maxwell."*

41 Into the 20th century

The pioneering 19th century work of scientists, engineers and medics in Scotland provided much of the foundation for the whirlwind of technological and scientific revolutions that characterised the 20th century. In those revolutions, Scotland's science would again be prominent. The legacies of Bain, Kelvin and Maxwell prompted the innovations in telephony and television of Alexander Graham Bell and John Logie Baird, and the radar inventions of Robert Watson-Watt. The antibiotic discoveries of Alexander Fleming would follow in the tradition of the great medical innovations in surgical antisepsis of Joseph Lister and the anaesthetic obstetrics of James Young Simpson. The brilliance and bravery of Mary Somerville, Elizabeth Garrett Anderson and the Edinburgh Seven broke the ceilings that enabled Agnes Blackadder and other women scientists to follow their careers of choice. The civil engineering developments of Rankine, Telford and McAdam would enable the proliferation of bridges, roads and railways that have transformed the transport systems of the modern world. And the discoveries in optical science of Brewster, Nicol and Kerr form part of the knowledge-base underpinning the creation of the fibre-optic networks that support the Internet and the World Wide Web. However, science is fundamentally a collaborative, incremental and international endeavour, and it has become more so. The collective, creative effort of large teams of scientists and engineers, drawn from across the world, is increasingly the best way to achieve rapid advances in science and technology. Within that context, the contributions of Scotland's scientists and technologists in the 20th and 21st centuries remained, and still remain, highly and perhaps disproportionately important. But that is another set of stories...

42 Sources and further reading

My principal research sources are listed chapter-by-chapter below. I have deliberately avoided the use of footnotes and specific detailed references, to keep the stories flowing and as accessible as possible for the general reader. This is not a reference work after all, but I hope that those interested in the subjects will find it easy and enjoyable to pursue their interests in the listed references. The resources of the World Wide Web must be mentioned again. On-line encyclopaedias these days provide an astoundingly comprehensive and generally accurate source of information. The amazing Wikipedia provides an excellent starting point for even the most obscure investigations, and I have rarely found it to be unreliable or inconsistent with other sources. Thank you Jimmy Wales. The huge efforts in digitising or scanning old books and scientific journals have created an on-line treasure-trove of information, available in a few mouse-clicks. The Internet Archive, JSTOR, Google, and the HathiTrust are just a few of the organisations who deserve huge credit. For those interested in more traditional sources, I can thoroughly recommend two highly readable books: 'Scotland's Story' by Tom Steel; and Arthur Henman's 'The Scottish Enlightenment - the Scots' Invention of the Modern World'.

1. Out of the dark ages
1. J.D. Mackie, 'A History of Scotland', pub. Penguin (1973)
2. Tom Steel, 'Scotland's Story', pub. Wm. Collins & Sons (1984)

2. The Logs and Bones of 'Marvellous Merchiston'
1. Mark Napier, 'Memoirs of John Napier of Merchiston', pub. Blackwell and Cadell (1834). Accessed at

https://books.google.co.uk/books?id=husGAAAAYAAJ&printsec=frontcover&dq=Memoirs+of+John+Napier+of+Merchiston
2. 'John Napier', Wikipedia. Accessed at https://en.wikipedia.org/wiki/John_Napier
3. 'John Napier', Famous scientists. Accessed at https://www.famousscientists.org/john-napier/
4. 'Napier's Bones, National Museum of Scotland. Accessed at https://www.nms.ac.uk/explore-our-collections/stories/science-and-technology/napiers-bones/
5. Picture credit Illustration 1: Unknown artist, 'John Napier, Inventor of Logarithms (1550-1617)' University of Edinburgh. Public domain {{PD-US-expired}} Accessed at https://commons.wikimedia.org/wiki/File:John_Napier.jpg {{PD-US-expired}}
6. Picture credit Illustration 2: Tiia Monto, Napier's 'bones' in Science Museum, London. Creative Commons licence https://creativecommons.org/licenses/by-sa/3.0/legalcode Accessed at https://commons.wikimedia.org/wiki/File:Napier%27s_rods.jpg

3. Scientist, freemason, soldier, spy - and founder of the Royal Society

1. Alexander Robertson, 'The Life of Sir Robert Moray, Soldier, Statesman and Man of Science ', pub. Longmans, Green & Co. (1922). Accessed at https://archive.org/details/lifeofsirrobertm00robeuoft
2. Robert Lomas, Gresham College lectures (2007) Accessed at http://www.gresham.ac.uk/login/?redirect_url=%2Flectures-and-events%2Fsir-robert-moray-soldier-scientist-spy-freemason-and-founder-of-the-royal
3. 'Sir Robert Moray', Wikipedia. Accessed at https://en.wikipedia.org/wiki/Robert_Moray
4. 'Robert Moray', The Royal Society, Science in the Making. Accessed at https://makingscience.royalsociety.org/s/rs/people/fst01643584

5. David Allan, 'Moray, Sir Robert', Oxford Dictionary of National Biography (2007). Accessed at https://doi.org/10.1093/ref:odnb/19645
6. Picture credit Illustration 3: Sir Robert Moray freemason mark. In a letter to Alexander Bruce, Moray wrote: *"This character or Hyeroglyphick, which I call a starre, is famous amongst the Egyptians and Grecians. For the Egyptian part of it I remitt you to Kircherus [Athanasius Kircher] bookes that I named in my last. The Greekes accounted it the symbol of health and tranquility of body and mind, as being composed of capitall letters that make up the word Hygieia, and I have applied five other letters to it that are the initials of 5 words that make up the summe of Christian Religion, as well as stoick philosophy, all which are to be found in it without much distortion or constraint, and make up the sweet word Agapa, which you know signifies love thou, or hee loves, which is the reciprocall love of God and man, and that same word is one of the 5 signified by the 5 letters. The rest are Gnothi, Pisteuei, Anecho, Apecho."* Accessed at https://freemasonry.bcy.ca/anti-masonry/pentagrams_additional.html

4. The creation of the pendulum clock

1. A.J. Youngson (1960). 'Alexander Bruce, F.R.S., Second Earl of Kincardine (1629-1681)'. Notes and Records of the Royal Society of London. The Royal Society. 15: 251–258. doi:10.1098/rsnr.1960.0024. Accessed at https://www.jstor.org/stable/531044
2. Broomhall blog, 'The clock that was 100 years too early'. Accessed at https://www.broomhallhouse.com/2018/12/the-clock-which-was-100-years-too-early/
3. Michael S. Mahoney, 'Christian Huygens: The Measurement of Time and of Longitude at Sea'. Published in Studies on Christiaan Huygens, pp 234-270, ed. H.J.M. Bos et al. (pub. Swets & Zeitlinger, 1980)
4. Dava Sobel, 'Longitude' pub. 4th Estate (1996)
5. Picture credit Illustration 4: Alexander Bruce by Johannes Mijtens. Accessed at https://commons.wikimedia.org/wiki/File:Alexander_Bruce_(1629-1680),_by_Johannes_Mijtens.jpg{{PD-US-expired}} .

6. Picture credit Illustration 5: The Bruce-Oosterwijck pendulum clock, Copyright National Museums of Scotland. Accessed at https://www.nms.ac.uk/explore-our-collections/stories/science-and-technology/bruce-oosterwijck-sea-clock/

5. Royal botanist and physician

1. 'Robert Morison', Edward Worth Library. Accessed at http://botany.edwardworthlibrary.ie/robert-morison/
2. Francis Wall Oliver, 'Makers of British Botany'. pub. Cambridge University Press (1913)
3. Scott Mandelbrote, 'Morison, Robert', Oxford Dictionary of National Biography (2004) Accessed at https://www.oxforddnb.com/view/10.1093/ref:odnb/9780198614128.001.0001/odnb-9780198614128-e-19275
4. Scott Mandelbrote, 'The Publication and Illustration of Robert Morison's Plantarum historiae universalis Oxoniensis', Huntington Library Quarterly Vol. 78, No.2, pp. 349-379 (2015). Accessed at https://www.jstor.org/stable/10.1525/hlq.2015.78.2.349
5. Picture credit Illustration 6: Woodcut of Robert Morison. Accessed at https://commons.wikimedia.org/wiki/File:Robert_Morison.png {{PD-US-expired}}

6. Telescopes and trigonometry

1. H. W. Turnbull, "The Tercentenary of the birth of James Gregory" (1938). Accessed at http://www-groups.dcs.st-and.ac.uk/~history/Extras/Turnbull_address.html
2. Niccolò Guicciardini, 'Gregory, James', Oxford Dictionary of National Biography (2004). Accessed at https://doi.org/10.1093/ref:odnb/11465
3. 'Gregorian telescope', Wikipedia. Accessed at https://en.wikipedia.org/wiki/Gregorian_telescope
4. Picture credit Illustration 7: James Gregory, by John Scougal. Accessed at https://commons.wikimedia.org/wiki/File:James_Gregory.jpeg {{PD-US-expired}}

7. The Scottish Enlightenment

1. Tom Steel, 'Scotland's Story', pub. Wm. Collins & Sons (1984)
2. Arthur Herman, 'The Scottish Enlightenment', pub. Fourth Estate (2003)
3. 'Education in early modern Scotland', Wikipedia, accessed at https://en.wikipedia.org/wiki/Education_in_early_modern_Scotland

8. The world's youngest professor

1. Charles Platts, 'Maclaurin, Colin', Oxford Dictionary of National Biography (1893). Accessed at https://doi.org/10.1093/odnb/9780192683120.013.17643
2. Erik Lars Sageng, 'Maclaurin, Colin', Oxford Dictionary of National Biography (2015). Accessed at https://doi.org/10.1093/ref:odnb/17643
3. Ian Tweddle, 'The prickly genius - Colin MacLaurin (1698-1746)' (1998). Accessed at https://web.archive.org/web/20080229073702/http://www.m-a.org.uk/docs/library/2064.pdf
4. Betty Ponting, 'Mathematics at Aberdeen', The Aberdeen University Review Vol. XLVIII (1980). Accessed at http://mathshistory.st-andrews.ac.uk/Extras/Aberdeen_1.html and http://www-history.mcs.st-andrews.ac.uk/Extras/Aberdeen_3.html
5. Charles Platts and George Molland, 'Liddel, Duncan', Oxford Dictionary of National Biography (2004). Accessed at https://doi.org/10.1093/ref:odnb/16639
6. Picture credit: Illustration 8: Colin Maclaurin woodcut. Accessed at https://www.nationalgalleries.org/art-and-artists/3143/colin-maclaurin-1698-1746-mathematician, Public Domain, https://commons.wikimedia.org/w/index.php?curid=1672996{{PD-US-expired}}

9. Hume's Fork and Humanism

1. 'David Hume (1711-1776)' The Internet Encyclopedia of Philosophy. Accessed at https://www.iep.utm.edu/hume/

2. Arthur Henman, 'The Scottish Enlightenment' pub. 4th Estate (2003)
3. Amyas Merivale, "How David Hume became the First Modern Humanist', Conway Hall Lecture (2011). Accessed at https://conwayhall.org.uk/ethicalrecord/david-hume-became-first-modern-humanist/
4. Ted Morris, 'David Hume's Life and Works', The Hume Society (2009). Accessed at https://web.archive.org/web/20180401123150/http://www.humesociety.org/about/HumeBiography.asp
5. William Edward Morris and Charlotte R. Brown, 'David Hume' Stanford Encyclopedia of Philosophy. Accessed at https://plato.stanford.edu/entries/hume/#pagetopright
6. 'David Hume' at Undiscovered Scotland. Accessed at https://www.undiscoveredscotland.co.uk/usbiography/h/davidhume.html
7. David Hume, 'My Own Life', (originally pub.1777) pub. Cosimo Classics (2015)
8. 'Hume Texts Online'. Accessed at https://davidhume.org/
9. Picture credit Illustration 9: Portrait of David Hume by Allan Ramsay, Scottish National Gallery. Accessed at https://commons.wikimedia.org/wiki/File:Allan_Ramsay_-_David_Hume,_1711_-_1776._Historian_and_philosopher_-_Google_Art_Project.jpg {{PD-US-expired}}

10. Founding the science of economics

1. 'About Adam Smith' Adam Smith Institute. Accessed at https://www.adamsmith.org/about-adam-smith
2. Tejvan Pettinger, "Biography of Adam Smith", Oxford, UK (2009) Accessed at https://www.biographyonline.net/writers/adam-smith.html
3. 'Biography of Adam Smith' Accessed at http://www.let.rug.nl/usa/biographies/adam-smith/
4. Robert L. Heilbroner, 'The Essential Adam Smith', Pub Norton & Co. (1986)
5. Picture credit Illustration 10, Adam Smith after a portrait by James Tassie. Accessed at

https://commons.wikimedia.org/wiki/File:AdamSmith.jpg {{PD-US-expired}}

11. Anatomy, obstetrics and scientific surgery

1. 'Dr William Hunter', The Hunterian, Accessed at https://www.gla.ac.uk/hunterian/about/history/drwilliamhunter
2. Jorge Dagnino-Sepúlveda, 'William Hunter (1718-1783): his legacy three hundred years from his birthday', Rev. Med. Chile vol.147 no.1 Santiago (2019) Accessed at https://scielo.conicyt.cl/scielo.php?script=sci_arttext&pid=S0034-9887201900010009 (original in Spanish)
3. Wendy Moore, 'The Knife Man: the Extraordinary Life and Times of John Hunter, Father of Modern Surgery', pub. Bantam (2005) reviewed by P.D. Smith at https://www.theguardian.com/books/2005/mar/12/featuresreviews.guardianreview8
4. 'John Hunter', Whonamedit, Accessed at http://www.whonamedit.com/doctor.cfm/84.html
5. Picture credit Illustration 11: William Hunter by Allan Ramsay, Hunterian Museum and Art Gallery. Accessed at https://commons.wikimedia.org/wiki/File:William_Hunter_(anatomist).jpg
6. Picture credit Illustration 12: John Hunter by Sir Joshua Reynolds, Wellcome Collection. Accessed at https://wellcomecollection.org/works/dp4t6euu under Creative Commons attribution CC BY 4.0
7. Picture credit Illustration 13: Anatomy of late pregnancy by Jan van Rymsdyk from 'Anatomy of the Human Gravid Uterus', W. Hunter (1774). Accessed at https://commons.wikimedia.org/wiki/File:Pregnancy_by_Jan_van_Riemsdyk_and_William_Hunter.jpg {{PD-US-expired}}

12. Lemons, Limeys, and Scots against scurvy

1. Peter M Dunn, 'James Lind (1716-94) of Edinburgh and the treatment of scurvy', Archives of Disease in Childhood 1997;76:F64–F65, (1997) Accessed at https://www.ncbi.nlm.nih.gov/pmc/articles/PMC1720613/

2. 'James Lind (1716 - 1794)', BBC History, Accessed at http://www.bbc.co.uk/history/historic_figures/lind_james.shtml
3. Simon Singh and Edzard Ernst, 'Trick or Treatment?', pub. Corgi (2009)
4. Sir Iain Chalmers, 'James Lind's legacy', James Lind Institute. Accessed at http://www.jli.edu.in/about/james-linds-legacy/
5. The James Lind Library. Accessed at https://www.jameslindlibrary.org/
6. David Harvie, 'Limeys - the Conquest of Scurvy', pub. Sutton (2005)
7. Mary Wharton, 'Sir Gilbert Blane Bt (1749-1834)', Annals of the Royal College of Surgeons of England, **66** (5): 375–6 (1984) Accessed at https://www.ncbi.nlm.nih.gov/pmc/articles/PMC2493700/
8. A. W. Beasley 'Sir Gilbert Blane Bt (1749-1834)', Annals of the Royal College of Surgeons of England, **67** (5): 332–3 (1985) Accessed at https://www.ncbi.nlm.nih.gov/pmc/articles/PMC2499530/
9. J. Wallace, 'Blane, Sir Gilbert, first baronet', Oxford Dictionary of National Biography (2004). Accessed at https://doi.org/10.1093/ref:odnb/2621
10. Surg Lt Cdr Jowan G Penn-Barwell, 'Sir Gilbert Blane FRS: the man and his legacy', J Royal Naval Medical Service, Vol 102.1, pp 61- 66 (2016). Accessed at https://web.archive.org/web/20160822031114/http://www.jrnms.com/wp-content/uploads/2016/07/JRNMS-102-1-61-66.pdf
11. 'Scurvy', Wikipedia. Accessed at https://en.wikipedia.org/wiki/Scurvy
12. 'Rose's Lime Juice', Wikipedia. Accessed at https://en.wikipedia.org/wiki/Rose%27s_lime_juice
13. 'Quick-selling Lime: L. Rose & Co.', Let's Look Again (2016). Accessed at http://letslookagain.com/tag/lauchlan-rose/
14. 'Roses Lime Juice', Leith Local History Society (2018) Accessed at http://www.leithlocalhistorysociety.org.uk/businesses/roses_lime_juice.htm
15. Picture credit Illustration 14: 'James Lind' by Sir George Chalmers. Public domain {{PD-US-expired}} Accessed at

https://commons.wikimedia.org/wiki/File:James_Lind_by_Chalmers.jpg_{{PD-US-expired}}_

13. Natural philosopher and natural educator

1. Paul Wood, 'Anderson, John', Oxford Dictionary of National Biography, (2004). Accessed at https://doi.org/10.1093/ref:odnb/481
2. Significant Scots, 'Anderson, John F.R.S', Significant Scots, Accessed at https://www.electricscotland.com/history/other/anderson_johnFRS.htm
3. 'John Anderson', History of the University of Glasgow. Accessed at https://universitystory.gla.ac.uk/biography/?id=WH0179&type=P
4. 'John Anderson', Research Resources in the University of Glasgow for Adam Smith and the Scottish Enlightenment. Accessed at https://www.gla.ac.uk/media/Media_260282_smxx.pdf
5. 'The Andersonian - the First Technical Institute' (2009). Accessed at https://technicaleducationmatters.org/2009/10/11/the-andersonian-the-first-technical-college
6. John Butt, 'John Anderson's Legacy - The University of Strathclyde and its Antecedents 1796-1996', pub. Tuckwell Press (1996)
7. Picture credit Illustration 15: John Anderson FRS, engraving by William Holl the Younger. Accessed at https://commons.wikimedia.org/wiki/File:John_Anderson_(zoologist).jpg {{PD-US-expired}}

14. The genesis of geology

1. National Library of Scotland, 'James Hutton 1726-1797', Scottish Science Hall of Fame, Accessed at https://digital.nls.uk/scientists/biographies/james-hutton/
2. Significant Scots, 'James Hutton' Accessed at https://docs.google.com/document/d/1iXi5LOJdtopPjqTmTnq8re0cYQGcHq_6gJRPiB-WSUE/edit#
3. John Playfair, 'Illustrations of the Huttonian Theory of the Earth', pub. Cadell and Davies (1802). Accessed at https://archive.org/details/NHM104643

4. Jean Jones, 'Hutton, James', Oxford Dictionary of National Biography (2013). Accessed at https://doi.org/10.1093/ref:odnb/14304
5. Simon Winchester, 'The Map that Changed the World', pub. Penguin (2002)
6. Picture credit, Illustration 16: James Hutton by Abner Lowe. Accessed at https://commons.wikimedia.org/wiki/File:James_Hutton.jpg {{PD-US-expired}}

15. Shedding light on heat, and the 'discovery' of air

1. Joseph Black, 'Experiments upon Magnesia Alba, Quick-Lime, and some other Alkaline Substances' (1756). Accessed at http://web.lemoyne.edu/~giunta/black.html
2. Sir William Ramsey, 'The Life and Letters of Joseph Black M.D.' (1918). Accessed at https://archive.org/details/lifelettersofjos00ramsrich/page/n7
3. Robert Chambers and Thomas Napier (Eds), 'A biographical dictionary of eminent Scotsmen/Black, Joseph', Accessed at https://en.wikisource.org/wiki/A_biographical_dictionary_of_eminent_Scotsmen/Black,_Joseph
4. Agnes Mary Clerke, 'Joseph Black', Dictionary of National Biography Vol. 05, Accessed at https://en.wikisource.org/wiki/Black,_Joseph_(DNB00)
5. John Robertson, 'The Scottish Enlightenment and the Militia Issue', pub. John Donald (1985) Accessed at https://archive.org/details/scottishenlighte0000robe
6. Bernard Barham Woodward, 'Daniel Rutherford', Oxford Dictionary of National Biography. Accessed at https://en.wikisource.org/wiki/Rutherford,_Daniel_(DNB00)
7. Daniel Rutherford, 'On the Air called Fixed, or Mephitic', doctoral dissertation, U. of Edinburgh (1772) translated by Crum Brown in J. Chem. Educ.1935,12,8, p370 (1935) . Accessed at https://pubs.acs.org/doi/10.1021/ed012p370
8. 'Daniel Rutherford 1749-1819', Royal College of Physicians of Edinburgh. Accessed at https://www.rcpe.ac.uk/heritage/art/rutherford-daniel-1749-1819

9. Picture credit, Illustration 17: Plaque of Joseph Black by James Tassie, Hunterian Museum Glasgow. Accessed at https://commons.wikimedia.org/wiki/File:Joseph_Black_plaque_by_James_Tassie,_Hunterian_Museum,_Glasgow.jpg under Creative Commons licence https://creativecommons.org/licenses/by-sa/4.0/deed.en {{PD-US-expired}}

16. Steam power efficiency

1. Jennifer Tann, 'Watt, James (1736-1819)', Oxford Dictionary of National Biography (2014). Accessed at https://www.oxforddnb.com/view/10.1093/ref:odnb/9780198614128.001.0001/odnb-9780198614128-e-28880
2. David Philip Miller, 'The Life and Legend of James Watt', pub. Pittsburgh (2019)
3. Ben Marsden, 'Watt's Perfect Engine', pub. Icon Books (2004)
4. M. Arago, 'Memoir on the Life of James Watt', Presentation to the French Academy of Sciences, pub. Edinburgh New Philosophical Journal Vol. 27, pp 221- 278 (1839) Accessed at https://archive.org/details/edinburghnewphil27edin/page/220
5. James Patrick Muirhead, 'The Life of James Watt: With Selections from His Correspondence', pub. John Murray (1858)
6. John Simkin, 'James Watt' (2016) Accessed at https://spartacus-educational.com/SCwatt.htm
7. 'James Watt 2019' Archives and Collections @ The Library of Birmingham. Accessed at https://theironroom.wordpress.com/2019/02/18/watt-2019-february/
8. Significant Scots 'James Watt', Accessed at https://electricscotland.com/history/men/james_watt.htm
9. Picture credit Illustration 18: James Watt - portrait by John Partridge after Sir William Beechley (1802). Accessed at https://commons.wikimedia.org/wiki/File:James-watt-1736-1819-engineer-inventor-of-the-stea.jpg {{PD-US-expired}}

17. Cerebral geniuses

1. 'Alexander Monro primus' Accessed at https://www.undiscoveredscotland.co.uk/usbiography/m/alexandermonro.html
2. Lisa Rosner, 'Monro, Alexander, secundus' Oxford Dictionary of National Biography (2004). Accessed at https://www.oxforddnb.com/view/10.1093/ref:odnb/9780198614128.001.0001/odnb-9780198614128-e-18965?docPos=4
3. R. Shane Tubbs, Peter Oakes, Ilavarasy S. Maran, Christian Salib, Marios Loukas, 'The foramen of Monro: a review of its anatomy, history, pathology, and surgery', Child's Nervous System, Volume 30, Issue 10, pp 1645–1649 (October 2014). Accessed at https://link.springer.com/article/10.1007/s00381-014-2512-6
4. Callum Maclean, 'George Kellie' Life in the Fast Lane, (2019). Accessed at https://litfl.com/george-kellie/
5. George Kellie, 'On Death from Cold and on Congestions of the Brain', Transactions of the Medico-Chirurgical Society of Edinburgh 1824;1:84-169 (1824). Accessed at https://archive.org/details/b22384315
6. Picture credit Illustration 19: Alexander Monro secundus - engraving by James Heath after Henry Raeburn (1800). U. of Otago Monro Collection. Accessed at https://commons.wikimedia.org/wiki/File:Alexander_Munro_secundus.jpg {{PD-US-expired}}

18. The Colossus of roads

1. Brenda J. Buchanan, 'McAdam, John Loudon (1756-1836)' (2004). Oxford Dictionary of National Biography doi:10.1093/ref:odnb/17325. Accessed at https://doi.org/10.1093/ref:odnb/17325
2. 'John Loudon McAdam', Significant Scots. Accessed at https://www.electricscotland.com/history/other/macadam_john.htm
3. Regina Jeffers, 'Regency Celebrity: John Loudon McAdam, Bringing Progress Through Road Improvements: Macadamisation', Every

Woman Dreams blog (2014). Accessed at https://reginajeffers.blog/2014/01/21/
4. John Loudon McAdam, 'A Practical Essay on the Scientific Repair and Preservation of Public Roads', pub. Neilson (1819). Accessed at https://archive.org/stream/cihm_47029#page/n4/mode/2up
5. John Loudon McAdam, 'Remarks on the Present System of Road Making', 9th ed. pub. Longmans et al., (1827). Accessed at https://books.google.co.uk/books?id=a9RMAAAAYAAJ&printsec=frontcover&dq=Remarks+on+the+present+system+of+road+making
6. 'Macadam' Wikipedia. Accessed at https://en.wikipedia.org/wiki/Macadam
7. Picture credit: Illustration 20: Charles Turner engraving of John Loudon Mcadam, British Museum. Accessed at https://commons.wikimedia.org/wiki/File:John_Loudon_McAdam.jpg {{PD-US-expired}}

19. Master of civil engineering

1. Roland Paxton, 'Telford, Thomas (1757-1834), Oxford Dictionary of National Biography (2013). Accessed at https://doi.org/10.1093/ref:odnb/27107
2. 'Thomas Telford (1757-1834), engineer whose works connected the United Kingdom, accelerating growth of trade and commerce', Scottish Engineering Hall of Fame. Accessed at http://www.engineeringhalloffame.org/profile-telford.html
3. John Rickman (Ed.), 'Life of Thomas Telford', pub. Hansard & Sons (1838). Accessed at https://archive.org/details/in.ernet.dli.2015.53284/
4. Julian Glover, 'Man of Iron - Thomas Telford and the Building of Britain", pub. Bloomsbury (2017)
5. L.T.C. Rolt, 'Thomas Telford', pub. Longmans, Green & Co. (1958)
6. Iainthepict, 'Thomas Telford' (2011). Accessed at http://iainthepict.blogspot.com/2011/08/thomas-telford.html
7. Picture credit: Illustration 21: Thomas Telford in 1838. Engraving by W. Raddon from a painting by S. Lake. Accessed at https://commons.wikimedia.org/wiki/File:ThomasTelford.jpg {{PD-US-expired}}

20. Rainwear from raintown

1. George Macintosh, 'Biographical Memoir of the Late Charles Macintosh, FRS' pub. Blackie & Co (1847). Accessed at https://books.google.co.uk/books?id=yd0AAAAAMAAJ&pg=PA152&dq=Charles+Macintosh&as_brr
2. R.B. Prosser and Geoffrey V. Morson, 'Macintosh, Charles', Oxford Dictionary of National Biography (2017). Accessed at https://doi.org/10.1093/ref:odnb/17541
3. 'The Patent Caoutchouc Case', The Mechanics Magazine, Vol. 24, pp 460-464 (1836). Accessed at https://books.google.co.uk/books?id=vA8FAAAAQAAJ&pg=PA460&lpg=PA460&dq=Everington+macintosh+court+case&source
4. Francis Espinasse and Ian Donnachie, 'Neilson, James Beaumont' Oxford Dictionary of National Biography (2009) Accessed at https://doi.org/10.1093/ref:odnb/19866
5. John Loadman, 'Tears of the Tree- the story of rubber - a modern marvel', pub. Oxford University Press (2005)
6. Picture credit Illustration 22: Charles Macintosh by Edward Burton from a portrait by John Graham Gilbert. Accessed at https://commons.wikimedia.org/wiki/File:Charles_Macintosh.jpg {{PD-US-expired}}
7. Picture credit Illustration 23: An advert for Macintosh rainwear in Carson, Pirie, Scott & Co's catalogue (1893, Chicago). Accessed at https://thevintageculture.com/storia-dell-impermeabile/ {{PD-US-expired}}

21. Enlightenment lights

1. Roland Paxton, 'Stevenson, Robert', Oxford Dictionary of National Biography (2004). Accessed at https://doi.org/10.1093/ref:odnb/26436
2. Northern Lighthouse Board, 'Our History', 'Stevenson engineers', and 'Lighthouses'. Accessed at http://nlb.org.uk
3. Eleanor Knowles, 'Robert Stevenson', Engineering Timelines. Accessed at http://www.engineering-timelines.com/who/Stevenson_R/stevensonRobert.asp

4. Christopher Spencer, 'Who built the Bell Rock Lighthouse?', BBC History (2017). Accessed at https://www.bbc.co.uk/history/british/empire_seapower/bell_rock_01.shtml
5. David Stevenson, 'Life of Robert Stevenson, Civil Engineer', pub. Adam & Charles Black (1878). Accessed at https://archive.org/details/lifeofrobertstev00stevrich/
6. Alan Stevenson, 'Biographical Sketch of the late Robert Stevenson', pub. Blackwood & Sons (1861). Accessed at https://play.google.com/books/reader?id=YSBGZ7ozsHMC&hl=en_GB&pg=GBS.PP8
7. Robert Stevenson, 'An Account of the Bell Rock Light-house', pub. Archibald Constable & Co. (1824). Accessed at http://www.gutenberg.org/files/48414/48414-h/48414-h.htm
8. Picture credit Illustration 24: Bust of Robert Stevenson, by Samuel Joseph. Accessed at https://commons.wikimedia.org/wiki/File:Robert_Stevenson_(lighthouse_engineer)_-_Google_Book_Search_-_Biographical_Sketch_of_the_Late_Robert_Stevenson.jpg {{PD-US-expired}}
9. Picture credit Illustration 25: J.M.W. Turner, 'Bell Rock Lighthouse during a storm from the North-East'. Accessed at https://commons.wikimedia.org/wiki/File:Joseph_Mallord_William_Turner_-_Bell_Rock_Lighthouse_-_Google_Art_Project.jpg {{PD-US-expired}}

22. Through a lens brightly

1. D.J. Mabberley, 'Brown, Robert', Oxford Dictionary of National Biography (2009). Accessed at https://doi.org/10.1093/ref:odnb/3645
2. Famous Scientists, 'Robert Brown' (2019). Accessed at https://www.famousscientists.org/robert-brown/
3. 'Matthew Flinders', Wikipedia (2019) Accessed at https://en.wikipedia.org/wiki/Matthew_Flinders
4. Amy Freeborn, Natural History Museum, 'How a seaweed scientist helped win the war' (2014). Accessed at https://www.nhm.ac.uk/natureplus/blogs/behind-the-

scenes/2014/03/26/how-a-seaweed-scientist-helped-win-the-war.html
5. Encyclopedia Britannica, 'Brownian Motion' (2019). Accessed at https://www.britannica.com/science/Brownian-motion
6. 'Brownian Motion' Wikipedia, (2019). Accessed at https://en.wikipedia.org/wiki/Brownian_motion
7. Robert Brown, 'The Miscellaneous Botanical Works of Robert Brown, Vol I ', pub. Hardwicke (1866). Accessed at https://archive.org/details/miscellaneousbot01brow/
8. Picture credit Illustration 26: Robert Brown by C. Fox after H.W. Pickersgill, Wellcome Trust. Accessed at https://commons.wikimedia.org/wiki/File:Portrait_of_Robert_Brown_Wellcome_L0011784.jpg under Creative Commons Attribution 4.0 International licence (CC BY 4.0) {{PD-US-expired}}

23. The Imperial impulse

1. 'Demographic history of Scotland', Wikipedia, Accessed at https://en.wikipedia.org/wiki/Demographic_history_of_Scotland (2020)
2. Tom Steel, 'Scotland's Story', pub. Wm. Collins & Sons (1984)
3. John Butt, 'John Anderson's Legacy - the University of Strathclyde and its Antecedents 1796-1996', pub. Tuckwell Press (1996)

24. The Queen of 19th century science

1. Mary R.S. Creese 'Somerville, Mary', Oxford Dictionary of National Biography (2004). Accessed at https://doi.org/10.1093/ref:odnb/26024
2. Elizabeth C Patterson, 'Mary Somerville and the Cultivation of Science 1815-1840', pub. Martinus Nijhoff (1983)
3. Martha Somerville (ed.), 'Personal Recollections from Early Life to Old Age of Mary Somerville', pub. John Murray (1873). Accessed at https://archive.org/details/b21960239
4. Mary Somerville, 'On the magnetizing power of the more refrangible solar rays', Philosophical Transactions of the Royal Society of London, Vol. 116, No. 1/3 pp. 132-139 (1826). Accessed at https://www.jstor.org/stable/107805

5. Skullsinthestars, 'A physics history-mystery: magnetism from light?' (2009). Accessed at https://skullsinthestars.com/2009/02/08/a-physics-history-mystery-magnetism-from-light/
6. Mary Somerville, 'On the Connexion of the Physical Sciences', 9th edition pub. John Murray (1858). Accessed at Project Gutenberg http://www.gutenberg.org/files/52869/52869-h/52869-h.htm
7. Mary Somerville, 'Physical Geography', American edition pub. Blanchard and Lea (1855). Accessed at https://archive.org/details/physicalgeograph00somerich
8. Picture credit Illustration 27: 'Mary Fairfax, Mrs William Somerville, 1780-1872' by Thomas Phillips (1834). Accessed at https://commons.wikimedia.org/wiki/File:Thomas_Phillips_-_Mary_Fairfax,_Mrs_William_Somerville,_1780_-_1872._Writer_on_science_-_Google_Art_Project.jpg {{PD-US-expired}}

25. Father of modern optics

1. Margaret Maria Gordon, 'The Home Life of Sir David Brewster', pub. David Douglas (1881). Accessed at https://archive.org/stream/homelifeofsirdav00gord
2. David Brewster, 'A Treatise on New Philosophical Instruments for Various Purposes in the Arts and Sciences'. Pub. John Murray and William Blackwood (1813). Accessed at https://archive.org/details/atreatiseonnewp00brewgoog/
3. David Brewster, 'On the polarization of Light by Oblique Transmission through All Bodies, Whether Crystallized or Uncrystallized', Philosophical Transactions of the Royal Society of London Vol. 104, pp. 219-230 (1814). Accessed at https://www.jstor.org/stable/107431
4. A.D. Morrison-Low, 'Brewster, Sir David', Oxford Dictionary of National Biography (2014). Accessed at https://doi.org/10.1093/ref:odnb/3371
5. Sir David Brewster, 'A Treatise on Optics', pub. Carey, Lea & Blanchard (1838). Accessed at http://lhldigital.lindahall.org/cdm/
6. Sir David Brewster, 'The Martyrs of Science', pub. John Murray (1841). Accessed at http://www.gutenberg.org/files/25992/25992-h/25992-h.htm

7. Sir David Brewster, 'More Worlds than One', pub. Robert Carter & Bros. (1856). Accessed at https://archive.org/details/moreworldsthanon00brewuoft/
8. David Brewster, 'A Treatise on the Kaleidoscope', pub. Archibald Constable (1819). Accessed at https://archive.org/details/b29295440/
9. Picture credit Illustration 28: Sir David Brewster with his Stereoscope, by D.J. Pound from a photograph by John Watkins. (Wellcome Library, London). Accessed at https://commons.wikimedia.org/wiki/File:Portrait_of_Sir_David_Brewster_Wellcome_M0018146.jpg under Creative Commons Attribution only licence CC BY 4.0

26. Pioneering polarizer

1. A.D. Morrison-Low, 'Nicol, William', Oxford Dictionary of National Biography (2006). Accessed at https://doi.org/10.1093/ref:odnb/37811
2. William Nicol, 'On a Method of so far increasing the Divergency of the two Rays in Calcareous-spar, that only one Image may be seen at a time', Edinburgh New Philosophical Journal, vol. 6, pp 83–4, (1829). Accessed at https://www.biodiversitylibrary.org/item/20135#page/103/mode/1up
3. Andrea Sella, University College London, 'Nicol's Prism', online article (2016). Accessed at https://www.chemistryworld.com/opinion/nicols-prism/1010178.article
4. A.D. Morrison-Low, 'Moyes, Henry', Oxford Dictionary of National Biography (2006). Accessed at https://doi.org/10.1093/ref:odnb/53551
5. H.F. Talbot, 'Facts relating to Optical Science no. II: On Mr. Nicol's Polarizing Eye-Piece', London and Edinburgh Philosophical Magazine, 3rd ser., vol. 4, no. 22, pp 289-90 (1834). Accessed at https://books.google.co.uk/books?id=pkEwAAAAIAAJ&pg=PA289&dq=Facts+relating+to+optical+science,+no.+II:+on+Mr+Nicol%27s+polarizing+eye-piece&hl

6. Anonymous, 'Means to enlarge the divergence of the two images of a calcite so that only one is seen at a time', (in German), Annalen der Physik und Chemie, band 29, no. 9, Part 1, paper 21, pp 182-186 (1833). Accessed at https://gallica.bnf.fr/ark:/12148/bpt6k15114b/f192.item.r=nicol
7. J.T. Silbermann, 'Search for an explanation of the Tufts, that are visible with the naked eye in polarized light', (in German), Annalen der Physik und Chemie, vol. 70, pp 393-399 (1847). Accessed at https://books.google.co.uk/books?id=2g0AAAAMAAJ&pg=PA393&lpg=PA393&dq
8. Wilhelm Haidinger, 'On the direct perception of polarized light and its plane of polarization' (in German), Annalen der Physik und Chemie, vol. 139, Issue 9, pp 29-39 (1844). Accessed at https://gallica.bnf.fr/ark:/12148/bpt6k15148n/f39.item
9. Picture credit Illustration 29: 'Henry Moyes and William Nicol' (1806) W. Ward mezzotint after portrait by J. Reubens Smith. (Wellcome Library, London). Accessed at https://commons.wikimedia.org/wiki/File:Henry_Moyes._Mezzotint_by_W._Ward,_1806,_after_J._R._Smith._Wellcome_V0006565.jpg under Creative Commons Attribution only licence CC BY 4.0 {{PD-US-expired}}
10. Picture credit Illustration 30: A diagram of an example of a Nicol prism, by Fred the Oyster. Accessed at https://commons.wikimedia.org/wiki/File:Nicol_prism.svg under the Creative Commons Attribution-Share Alike 4.0 International licence.

27. The Stirling Engine

1. Significant Scots, 'Robert Stirling'. Accessed at https://www.electricscotland.com/history/men/stirling_robert.htm (2019)
2. Andy Ross, 'Stirling Cycle Engines', pub. Solar Engines (1977)
3. Wikipedia, 'Stirling Engine'. Accessed at https://en.wikipedia.org/wiki/Stirling_engine (2019)
4. Scottish Engineering Hall of Fame, 'Reverend Doctor Robert Stirling (1790-1878)'. Accessed at http://www.engineeringhalloffame.org/profile-stirling.html (2019)

5. John S Reid, (U. of Aberdeen), 'Stirling Stuff'. Accessed at https://arxiv.org/ftp/arxiv/papers/1604/1604.02362.pdf (2019)
6. 'The Air Engine' in The Engineer, Dec. 14th p 516 (1917). Accessed at http://hotairengines.org/stirling-engines-inventors/stirling/the-stirling-engine-of-1816/the-air-engine (2019)
7. Ben Marsden, 'Stirling, Robert', Oxford Dictionary of National Biography, (2004). Accessed at https://doi.org/10.1093/ref:odnb/26534
8. James Stirling, 'The Stirling Engine', Proc. Inst. Civil Engineers, Vol. IV, p.348 Part 1 (1845). Accessed at http://hotairengines.org/stirling-engines-inventors/stirling/the-stirling-engine-of-1842/complete-description
9. Wikipedia, 'Applications of the Stirling Engine'. Accessed at https://en.wikipedia.org/wiki/Applications_of_the_Stirling_engine (2019)
10. Cryogenic Engineering Group, U. of Oxford. Accessed at http://www2.eng.ox.ac.uk/cryogenics/research/cryocoolers-for-space-applications (2019)
11. Picture credit Illustration 31: Stirling's Economiser and Hot Air Engine. From his patent no. 4081 of 1816. Accessed at http://hotairengines.org/stirling-engines-inventors/stirling/the-stirling-engine-of-1816/the-air-engine (2020) {{PD-US-expired}}

28. Anatomy at the edge of legality

1. Henry Lonsdale, 'A Sketch of the Life and Writings of Robert Knox', pub. Macmillan (1870). Accessed at https://archive.org/details/asketchlifeandw01lonsgoog
2. Claire L. Taylor, 'Knox, Robert', Oxford Dictionary of National Biography (2004). Accessed at https://doi.org/10.1093/ref:odnb/15787
3. George Thomas Bettany, 'Knox, Robert', Dictionary of National Biography, 1885-1900, Vol. 31. Accessed at https://en.wikisource.org/wiki/Knox,_Robert_(1791-1862)_(DNB00)

4. Lisa Rosner, 'The Anatomy Murders', pub. U. of Pennsylvania Press (2010). Accessed at https://archive.org/details/anatomymurdersbe00rosn
5. Andrew S. Currie, 'Robert Knox, Anatomist, Scientist and Martyr', Proc. Royal Society of Medicine, Vol. 26, pp 39-46 (1932). Accessed at https://journals.sagepub.com/doi/pdf/10.1177/003591573202600111
6. I. Maclaren, 'Robert Knox MD, FRCSEd, FRSEd 1791-1862: The first Conservator of the College Museum', J.R. Coll.Surg. Edinb., Vol. 45, 392-397 (2000). Accessed at https://web.archive.org/web/20070310154552/http://www.rcsed.ac.uk/journal/vol45_6/4560011.htm
7. A.W. Bates, 'The Anatomy of Robert Knox', pub. Sussex Academic press (2010)
8. Robert Knox, 'On the Climate of Southern Africa…', Edinburgh Philosophical Journal, vol. 5, no. 10, pp 279-286 (1821). Accessed at https://books.google.co.uk/books?id=-hYxAQAAMAAJ&pg=PA236-IA1&lpg=PA236-IA1&dq
9. Robert Knox, 'Observations on the Comparative Anatomy of the Eye', Trans. Royal Society of Edinburgh, Vol. 10 (1823). Accessed at https://books.google.co.uk/books?id=aqNbAAAAcAAJ&pg=PA43&lpg
10. Robert Knox, 'Great Artists and Great Anatomists', pub. John van Voorst (1852). Accessed at https://books.google.co.uk/books?id=tDgBAAAAQAAJ&pg=PR12&dq=Great+artists+and+great+anatomists&hl
11. Robert Knox, 'The Races of Men: a Fragment', pub. Lea & Blanchard (1850). Accessed at https://archive.org/details/bub_gb_XwQXAAAAYAAJ
12. Andries Stockenstrom, 'The Autobiography of Sir Andries Stockenstrom', pub. Juta & Co. (1887). Accessed at https://archive.org/details/autobiographyofl01stoc
13. Picture credit Illustration 32: Dr. Robert Knox, surgeon and anatomist. Artist unknown, public domain. Source Lonsdale, Henry (1870), 'A Sketch of the Life and Writings of Robert Knox the Anatomist', London: MacMillan and Co. Accessed at

https://commons.wikimedia.org/wiki/File:Dr._Robert_Knox.jpg
{{PD-US-expired}}

29. Geological time and the antiquity of Man

1. Martin Rudwick, 'Lyell, Sir Charles, first baronet', Oxford Dictionary of National Biography (2012). Accessed at https://doi.org/10.1093/ref:odnb/17243
2. Charles Lyell, 'Life, letters and journals of Sir Charles Lyell, Bart', ed. Katherine M Lyell, pub. John Murray (1881). Accessed at https://archive.org/details/cu31924012129544
3. Charles Lyell, 'Principles of Geology', Vol I, pub. John Murray (1830). Accessed at https://archive.org/details/PrinciplesgeoloVol1Lyel
4. Charles Lyell, 'Principles of Geology', Vol II, Tenth edition, pub. John Murray (1868). Accessed at https://archive.org/details/NHM64258B
5. Charles Lyall, 'Elements of Geology', pub. John Murray (1838). Accessed at https://archive.org/details/elementsgeology00lyelgoog
6. Charles Lyell, 'The geological evidences of the antiquity of Man', pub. J.B. Lippincott (1871). Accessed at https://archive.org/details/60411780R.nlm.nih.gov
7. M.J.S. Rudwick, 'Lyell and the Principles of Geology', in: D.J. Blundell & A.C. Scott, (eds) 'Lyell: the Past is the Key to the Present'. Geological Society, London, Special Publications, vol.143, 3-15. Accessed at https://sp.lyellcollection.org/content/specpubgsl/143/1/1.full.pdf
8. Bill Bryson, 'A Short History of Nearly Everything', pp 90-108, pub. Black Swan (2004)
9. University of Cambridge, 'Darwin Correspondence Project'. Accessed (2019) at 'Charles Lyell' https://www.darwinproject.ac.uk/charles-lyell and letters https://www.darwinproject.ac.uk/letter/DCP-LETT-1870.xml and https://www.darwinproject.ac.uk/letter/DCP-LETT-4035.xml
10. Picture credit Illustration 33: Charles Lyell, geologist (1797-1875). Unknown author. Accessed at https://commons.wikimedia.org/wiki/File:Charles_Lyell00.jpg
{{PD-US-expired}}

30. Electric clocks and electric thoughts

1. Ivan S. Ruddock, 'Alexander Bain: The real father of television?', Scottish Local History, Iss. 83 (2012). Accessed at https://www.slhf.org/sites/default/files/publications/slhf12_alexanderbain.pdf
2. R.W. Burns, 'Bain, Alexander', Oxford Dictionary of National Biography (2004). Accessed at https://doi.org/10.1093/ref:odnb/1080
3. John Finlaison, 'An account of some remarkable applications of the electric fluid to the useful arts, by Mr. Alexander Bain; with a vindication of his claim to be the first inventor of the electro-magnetic printing telegraph, and also of the electro-magnetic clock' pub. Chapman and Hall (1843). Accessed at https://archive.org/details/accountsomerema00Finl/page/n8/mode/2up
4. Alexander Bain, 'A short history of the electric clocks', pub. Chapman and Hall (1852). Accessed at https://books.google.co.uk/books/about/A_short_history_of_the_electric_clocks.html
5. Steven Roberts, 'Distant Writing, A history of the telegraph companies in Britain 1838-1868: Bain' (2012). Accessed at https://distantwriting.co.uk/bain.html
6. A.G. Thomson, Royal Scottish Museum, 'The First Electric Clock: Alexander Bain's gold contact system". Accessed at https://core.ac.uk/download/pdf/81100144.pdf (2020)
7. 'Bain, the Inventor of the Chemical Telegraph', Scientific American, vol. 8, no 33, p 258, (1853). Accessed at https://www.scientificamerican.com/article/bain-the-inventor-of-the-chemical-t
8. Picture credit Illustration 34: Alexander Bain, unknown artist c.1876. Accessed at https://commons.wikimedia.org/wiki/File:Alexander_Bain.jpg {{PD-US-expired}}
9. Picture credit Illustration 35: Alexander Bain's improved facsimile 1850, European Patent Office. Accessed at https://commons.wikimedia.org/w/index.php?curid=7277453 {{PD-US-expired}}

31. Diffusion and dialysis

1. Michael Stanley, 'Graham, Thomas' Oxford Dictionary of National Biography (2008). Accessed at https://doi.org/10.1093/ref:odnb/11224
2. J. Stewart Cameron, 'Thomas Graham (1805-1869) - The "Father" of Dialysis', in 'Dialysis: History, Development and Promise', Chap. 1C, pp 19-25, pub. World Scientific (2012). Accessed at https://books.google.co.uk/books?id=nv3R3U3kwO8C&printsec
3. Undiscovered Scotland, 'Thomas Graham'. Accessed at https://www.undiscoveredscotland.co.uk/usbiography/g/thomasgraham.html (2019)
4. Thomas Graham, 'On the Absorption of Gases by Liquids', Annals of Philosophy, Vol. 12 pp 69-74 (1826). Accessed at https://www.biodiversitylibrary.org/item/20149
5. 'Graham's Law', Wikipedia. Accessed at https://en.wikipedia.org/wiki/Graham%27s_law (2019)
6. Thomas Graham, 'The Bakerian Lecture - On Osmotic Force', Phil. Trans. Roy. Soc. Vol. 144, pp 177-228, (1854). Accessed at https://royalsocietypublishing.org/doi/pdf/10.1098/rstl.1854.0008
7. Thomas Graham, 'Liquid Diffusion Applied to Analysis', Phil. Trans. Roy. Soc. Vol. 151, pp 183-224 (1861). Accessed at https://archive.org/details/philtrans07144381
8. Jaime Wisniak, 'Thomas Graham, part I, Contributions to thermodynamics, chemistry, and the occlusion of gases', Educación Química, vol. 24, Iss. 3, pp 316-325 (2013). Accessed at https://www.sciencedirect.com/science/article/pii/S0187893X13724819
9. Jaime Wisniak, 'Thomas Graham, part II, Contributions to diffusion of gases and liquids, colloids, dialysis and osmosis', Educación Química, vol. 24, Supp. 2, pp 506-515 (2013). Accessed at https://www.sciencedirect.com/science/article/pii/S0187893X13725217
10. Picture credit Illustration 36: Thomas Graham, lithograph by Rudolph Hoffman (1856) after photograph by Beard of London. Accessed at https://commons.wikimedia.org/wiki/File:Thomas_Graham_Litho.JPG {{PD-US-expired}}

32. Great ships and solitary waves

1. Grace's Guide to British Industrial History : 'John Scott Russell Obituaries' Accessed at https://www.gracesguide.co.uk/John_Scott_Russell:_Obituaries
2. David K. Brown, 'Russell, John Scott', Oxford Dictionary of National Biography (2012). Accessed at https://doi.org/10.1093/ref:odnb/24328
3. Henry Petroski, 'John Scott Russell', American Scientist, Vol. 86, pp 18-21 (1998). Accessed at https://www.jstor.org/stable/27856932
4. Steven Brindle, 'Brunel: the Man who Built the World' pub. Weidenfeld & Nicolson (2005)
5. John Scott Russell, 'The Wave of Translation in the Oceans of Water, Air and Ether', pub. Trübner & Co. (1885). Accessed at https://archive.org/details/wavetranslation01russgoog
6. Larrie Ferreiro and Alexander Pollara, 'Clippers, yachts and the false promise of the wave line', Physics Today, Vol. 70, 7, pp 52-58 (2017). Accessed at https://physicstoday.scitation.org/doi/pdf/10.1063/PT.3.3627
7. Heriot-Watt University Dept. of Mathematics, 'John Scott Russell and the solitary wave'. Accessed at http://www.macs.hw.ac.uk/~chris/scott_russell.html (2019)
8. Heriot-Watt University, 'John Scott Russell's Soliton Wave Re-created' Accessed at http://www.ma.hw.ac.uk/solitons/press.html (2019)
9. M. Nakazawa et al., '80 Gbit/s multi-channel soliton transmission over transoceanic distances', in Akira Hasegawa (ed.) 'Massive WDM and TDM Soliton Transmission Systems: a ROSC Symposium', pub. Kluwer Academic Publishers (2002)
10. Monica Landgraf, 'Optical communication at record-high speed via soliton frequency combs generated in optical microresonators', at Phys.org (2017). Accessed at https://phys.org/news/2017-06-optical-record-high-soliton-frequency-microresonators.html
11. Theodore Hänsch, 'Passion for Precision', Nobel Lecture (2005) Accessed at https://www.nobelprize.org/uploads/2018/06/hansch-lecture.pdf
12. Picture credit Illustration 37: John Scott Russell, photographer unknown (1847). Accessed at

https://commons.wikimedia.org/wiki/File:Russell_J_Scott.jpg
{{PD-US-expired}}

33. Light from oil: James 'Paraffin' Young

1. National Candle Association, 'History of candles'. Accessed at http://www.candles.org/about_history.html (January 2020)
2. Michael Faraday, 'The chemical history of a candle', Royal Institution Lectures, (1848) ed. William Crookes. Accessed at http://www.gutenberg.org/cache/epub/14474/pg14474-images.html
3. John Butt, 'Young, James', Oxford Dictionary of National Biography (2004). Accessed at https://doi.org/10.1093/ref:odnb/30266
4. John Butt, 'James Young, Scottish Industrialist and Philanthropist', Ph.D. thesis, U. of Glasgow (1963). Accessed at http://theses.gla.ac.uk/3894/1/1963ButtPhD.pdf
5. F.M. Cook, 'What the oil industry owes to Dr. James Young', presented at the centenary of the Young Chair of Chemical Technology at the U. of Strathclyde (1970) reprinted from Chemistry and Industry (May 29 1971). Accessed at https://www.scottishshale.co.uk/DigitalAssets/pdf/BP/112486-002.pdf
6. Robert H. Kargon, 'Science in Victorian Manchester', pp 86-96 pub. Transaction Publishers (2010)
7. Baron Dupin, 'The Commercial Power of Great Britain; a complete view of the public works of this country', (translated from the French) vol. 2, pp 235-236 pub. Charles Knight (1825). Accessed at https://archive.org/details/commercialpowero02dupiuoft
8. 'Paraffin Young: Pioneer of Oil'. Biographical film documentary. Accessed at https://scotlandonscreen.org.uk/browse-films/007-000-000-031-c
9. Picture credit Illustration 38: James Young, Scottish chemist, photo by T. & R. Annan & Sons (1906). Frontispiece of *Bibliotheca Chemica* (Volume 2), edited by John Ferguson. Public Domain {{PD-US-expired}}. Accessed at https://en.wikipedia.org/wiki/James_Young_(chemist)#/media/File:Young_James_chemist.jpg

34. Childbirth and anaesthesia

1. J. Y. Simpson, 'Notes on the Employment of the Inhalation of Sulphuric Ether in the Practice of Midwifery', Monthly Journal of Medical Science, vol. 7, no. 9, pp 721-728 (1847). Accessed at https://babel.hathitrust.org/cgi/pt?id=hvd.32044103093258&view=1up&seq=901
2. James Young Simpson, 'The obstetric memoirs and contributions of James Y. Simpson', vol 2, ed. W.O. Priestley and Horatio R. Storer, Pub. Adam and Charles Black (1856). Accessed at https://books.google.co.uk/books?id=0uE-AAAAcAAJ&printsec
3. Medical News, 'On Etherisation as a Means of procuring Insensibility to Pain', Monthly Journal of Medical Science, vol. 7 no. 9, pp 636 - 640 (1847). Accessed at https://babel.hathitrust.org/cgi/pt?id=hvd.32044103093258&view=1up&seq=798
4. H. Laing Gordon, 'Sir James Young Simpson and Chloroform', pub. T. Fisher Unwin (1897). Accessed at https://www.gutenberg.org/files/34128/34128-h/34128-h.htm
5. J.Y. Simpson, 'Account of a New Anaesthetic Agent (chloroform) as a substitute for sulphuric ether', pamphlet, pub. Sutherland and Knox, (1848). Accessed at https://wellcomelibrary.org/item/b20642817#?c
6. Malcolm Nicolson, 'Simpson, Sir James Young, first baronet', Oxford Dictionary of National Biography, (2004). Accessed at https://doi.org/10.1093/ref:odnb/25584
7. Eve Blantyre Simpson, 'Sir James Y Simpson' pub. Oliphant Anderson & Ferrier (1896). Accessed at https://electricscotland.com/history/other/sirjamesysimpson00sim puoft.pdf
8. John Duns, 'Memoir of Sir James Y. Simpson, Bart' pub. Edmonston & Douglas (1873). Accessed at https://books.google.co.uk/books?id=JjcaBbRfFYQC&pg=PT20&dq=Memoir+of+Sir+James+Y.+Simpson,+bart.&hl
9. Sir J.Y. Simpson, 'Our existing System of Hospitalism and its Effects. Part II', Edinburgh Medical J. vol 14, no. 12, pp1084 - 1115 (1869). Accessed at

https://www.ncbi.nlm.nih.gov/pmc/articles/PMC5327028/pdf/edinbmedj73951-0026.pdf

10. Gilbert J. Grant, Abraham H. Grant, Charles J. Lockwood, 'Simpson, Semmelweis, and Transformational Change', Obstetrics and Gynecology vol. 106, no. 2, pp 384-387 (2005). Accessed at https://pdfs.semanticscholar.org/8f97/b38d36724000cb880870c5b9227277232116.pdf

11. Donald Caton, 'The History of Obstetric Anesthesia' in 'Chestnut's Obstetric Anesthesia' ed. David H. Chestnut, Linda S. Polley, Cynthia A. Wong pub. Elsevier Saunders (2014)

12. Picture credit Illustration 39: 'Sir James Young Simpson', photograph by Bingham. Wellcome Images. Accessed at https://commons.wikimedia.org/wiki/File:James_Young_Simpson_2.jpg under the Creative Commons Attribution 4.0 International licence https://creativecommons.org/licenses/by/4.0/

13. Picture credit Illustration 40: Obstetrical forceps, J.Y. Simpson, 19th C. Wellcome Images. Accessed at https://commons.wikimedia.org/wiki/File:Obstetrical_forceps,_J.Y._Simpson,_19thC_Wellcome_L0006323.jpg under Creative Commons Attribution only licence CC BY 4.0 https://creativecommons.org/licenses/by/4.0/_{{PD-US-expired}}.

35. Antiseptic surgery

1. G.T. Wrench, 'Lord Lister - His Life and Work', pub. T. Fisher Unwin (1913). Accessed at https://archive.org/details/lordlisterhislif00wrenuoft

2. John Bankston, 'Joseph Lister and the Story of Antiseptics', pub. Mitchell Lane (2005). Accessed at https://archive.org/details/josephlisterstor00john

3. Dennis Pitt and Jean-Michel Aubin, 'Joseph Lister: father of modern surgery', Can. J. Surg, vol 55, no. 5, E8-E9, (2012). Accessed at https://dx.doi.org/10.1503%2Fcjs.007112

4. Christopher Lawrence, 'Liston, Joseph', Oxford Dictionary of National Biography (2004). Accessed at https://doi.org/10.1093/ref:odnb/34553

5. Joseph Lister, 'On a new method of treating compound fracture, abscess etc', The Lancet, vol i, pp 326, 357, 387, 507 (1867) reprinted in 'The collected papers of Joseph, Baron Lister', vol 2, pp 1- 36, pub. Frowde, Hodder and Stoughton by the Clarendon Press (1909). Accessed at https://archive.org/details/collectedpaperso02listuoft
6. Joseph Lister, 'On the effects of the antiseptic system of treatment upon the salubrity of a surgical hospital', The Lancet, vol i, pp 4-40 (1870) reprinted in 'The collected papers of Joseph, Baron Lister', vol 2, pp 123-136, pub. Frowde, Hodder and Stoughton by the Clarendon Press (1909). Accessed at https://archive.org/details/collectedpaperso02listuoft
7. Albert Einstein, speaking at the Royal Albert Hall (1933). Accessed at https://www.royalalberthall.com/about-the-hall/news/2013/october/3-october-1933-albert-einstein-speaks-at-the-hall (2020)
8. Picture credit Illustration 41: R. Chapman, from 'The picture of Glasgow or a Stranger's Guide' showing Glasgow Royal Infirmary c.1812. Accessed at https://commons.wikimedia.org/wiki/File:Glasgow_Royal_Infirmary,_c1812.jpg under Creative Commons Public Domain Mark 1 licence {{PD-US-expired}}
9. Picture credit Illustration 42: Joseph Lister, 1st Baron Lister c.1855. Photographer unknown. Accessed at https://commons.wikimedia.org/wiki/File:Joseph_Lister_c1855.jpg Creative Commons Public Domain Mark 1 licence {{PD-US-expired}}

36. Elizabeth Garrett Anderson and the Edinburgh Seven

1. Louisa Garrett Anderson, 'Elizabeth Garrett Anderson 1836-1917', pub. Faber and Faber (1939). Accessed at https://archive.org/details/elizabethgarrett0000ande/mode/2up
2. Laura Kelly, 'Elizabeth Garrett Anderson: early pioneer of women in medicine', The Lancet, vol 390, pp 2620-2621 (2017). Accessed at https://strathprints.strath.ac.uk/62860/1/Kelly_The_Lancet_2017

_Elizabeth_Garrett_Anderson_early_pioneer_of_women_in_medicine.pdf
3. M.A. Elston, 'Anderson, Elizabeth Garrett', Oxford Dictionary of National Biography (2017). Accessed at https://doi.org/10.1093/ref:odnb/30406
4. Elizabeth Garrett Anderson, Spartacus Educational. Accessed at https://spartacus-educational.com/WandersonE.htm (2020)
5. M.A. Elston, 'Edinburgh Seven', Oxford Dictionary of National Biography (2015). Accessed at https://doi.org/10.1093/ref:odnb/61136
6. Margaret Todd, 'The Life of Sophia Jex-Blake', pub. Macmillan & Co (1918). Accessed at https://archive.org/details/lifeofsophiajexb00toddiala/mode/2up
7. 'Edinburgh Seven', Wikipedia. Accessed at https://en.wikipedia.org/wiki/Edinburgh_Seven (2020)
8. Sophia Jex-Blake, 'Medical Women', pub. Oliphant, Anderson & Ferrier (1886). Accessed at https://archive.org/details/medicalwomenthes00jexb/mode/2up
9. 'First female medical students get degrees at last', BBC News. Accessed at https://www.bbc.co.uk/news/uk-scotland-edinburgh-east-fife-48885287 (2020)
10. Picture credit Illustration 43: Elizabeth Garrett Anderson, photo by Stanislav Walery, pub. Sampson Low & Co. (1889). National Portrait Gallery. Accessed at https://commons.wikimedia.org/wiki/File:Elizabeth_Garrett_Anderson.jpg under licence CC-PD-Mark/PD-Art (PD-old-70)
11. Picture credit Illustration 44: Sophia Jex-Blake as a young woman. Unknown author, public domain. Accessed at https://commons.wikimedia.org/wiki/File:Sophia_Jex-Blake_as_a_young_woman.jpg under licence CC-PD-Mark/PD-Art (PD-old-70)

37. Magneto- and electro-optics

1. John Kerr, 'On the rotation of the Plane of Polarization by Reflection from the Pole of a Magnet ', Philosophical Magazine, Series 5, vol. 3, no.19, pp 321-343 (1877). Accessed at

https://www.biodiversitylibrary.org/item/122159#page/335/mode/1up

2. P. Weinberger, 'John Kerr and his effects found in 1877 and 1878', Philosophical Magazine Letters, vol. 88, 12, pp 897-907 (2008). Accessed at https://web.archive.org/web/20110718214456/http://www.computational-nanoscience.de/Weinberger/Famous-Papers/PML-2008.pdf

3. John Kerr, 'A new relation between electricity and light: Dielectrified media birefringent'. Philosophical Magazine, Series 4, vol. 50, no. 332, pp 337–348 (1875). Accessed at https://www.biodiversitylibrary.org/item/120578#page/350/mode/1up.

4. Robert Steele and Anita McConnell, 'Kerr, John' Oxford Dictionary of National Biography (2004). Accessed at https://doi.org/10.1093/ref:odnb/34300

5. Robert C. Gray, 'The Rev. John Kerr F.R.S., Inventor of the Kerr Cell', Nature, vol. 136, no. 3433, pp 245-247 (1935). Accessed at https://www.nature.com/articles/136245a0.pdf

6. C.G.K. 'The Rev. Dr. John Kerr, F.R.S.', (obituary) Nature, vol. 76, no. 1979, pp 575-576 (1907). Accessed at https://doi.org/10.1038%2F076575a0

7. University of Strathclyde archives, 'Glasgow Free Church Training College'. Accessed at https://atom.lib.strath.ac.uk/glasgow-free-church-training-college

8. Picture credit Illustration 45: Rev. John Kerr photo by Thomas Annan c. 1860. (National Portrait Gallery) Accessed at https://commons.wikimedia.org/w/index.php?curid=8252455 {{PD-US-expired}}

38. The Rankine Cycle

1. Ben Marsden 'Rankine, (William John) Macquorn', Oxford Dictionary of National Biography (2004). Accessed at https://doi.org/10.1093/ref:odnb/23133

2. Alexander Macfarlane, 'Lectures on Ten British Physicists of the Nineteenth Century', pp 22-37, pub. Wiley (1919). Accessed at https://archive.org/details/lecturesontenbri00macfrich

3. 'William John Macquorn Rankine, engineer, polymath, educator and researcher. Pioneer of thermodynamics', Scottish Engineering Hall of Fame. Accessed at http://www.engineeringhalloffame.org/profile-rankine.html (2019)
4. W.J. Macquorn Rankine, 'Miscellaneous Papers', W.J. Millar (ed.) pub. Charles Griffin (1881). Accessed at https://archive.org/details/miscellaneoussci00rank
5. W.J. Macquorn Rankine, ' 'On the mechanical action of heat, especially in gases and vapours', Trans. Royal Soc. Edinburgh, vol. 20, part 1, pp 147- 193 (1850). Accessed at https://archive.org/details/transactionsofro20royal/page/146
6. W.J. Macquorn Rankine, 'On the centrifugal theory of elasticity, and its connection with the theory of heat', Trans. Royal Soc. Edinburgh, vol. 20, part 3, pp 425-440 (1850). Accessed at https://archive.org/details/transactionsofro20royal/page/424
7. Picture credit Illustration 46: 'W.J. Macquorn Rankine' by Thomas Annan. Accessed at https://commons.wikimedia.org/wiki/File:William_Rankine_1870s.jpg {{PD-US-expired}}

39. Absolute temperature and Thermodynamics

1. Silvanus P. Thompson, 'The Life of William Thomson, Baron Kelvin of Largs', vol. 1, pub. Macmillan & Co. (1910). Accessed at https://archive.org/details/b31360403_0001/
2. George F. Fitzgerald, 'Lord Kelvin, Professor of Natural Philosophy in the University of Glasgow', pub James MacLehose & Sons (1899). Accessed at https://www.electricscotland.com/history/other/lordkelvinprofes00fitzrich.pdf
3. Wikipedia, 'William Thomson, 1st Baron Kelvin' Accessed at https://en.wikipedia.org/wiki/William_Thomson,_1st_Baron_Kelvin
4. Crosbie Smith, 'Thomson, William, Baron Kelvin', Oxford Dictionary of National Biography (2011). Accessed at https://doi.org/10.1093/ref:odnb/36507
5. P.Q.R., "On Fourier's expansions of functions in trigonometric series" Cambridge Mathematical Journal vol.2, pp 258–262 (1841).

Accessed at https://babel.hathitrust.org/cgi/pt?id=nyp.33433062744911&view=1up&seq=274

6. P.Q.R., 'On the uniform motion of heat in homogeneous solid bodies, and its connection with the mathematical theory of electricity', Cambridge Mathematical Journal vol.3, pp 71-84 (1842). Accessed at https://babel.hathitrust.org/cgi/pt?id=nyp.33433062744929&view=1up&seq=85

7. William Thomson, 'Demonstration of a Fundamental Proposition in the Mechanical Theory of Electricity', Cambridge Mathematical Journal vol.4, pp 223-226 (1845). Accessed at https://babel.hathitrust.org/cgi/pt?id=nyp.33433062745009&view=1up&seq=247

8. N.L.S Carnot, 'Reflections on the motive power of heat and on machines fit to develop that power' (reprinted and translated from the original 1824 French paper by R.H. Thurston), pub. Wiley (1897). Accessed at https://www3.nd.edu/~powers/ame.20231/carnot1897.pdf

9. James P. Joule, 'On the existence of an Equivalent Relation between Heat and the ordinary Forms of Mechanical Power', Philosophical Magazine, Ser. 3, vol. 27 pp 205-207 (1845). Accessed at https://babel.hathitrust.org/cgi/pt?id=umn.319510006140796&view=1up&seq=225

10. William Thomson, 'On an Absolute Thermometric scale founded on Carnot's theory of the motive power of heat, and calculated from Regnault's observations', Proc. Cambridge Philosophical Society, (June 1848), reprinted in Mathematical and Physical Papers, vol. 1, pp 100-106 (1882). Accessed at https://archive.org/details/mathematicaland01kelvgoog/page/n124

11. William Thomson, 'An account of Carnot's theory of the motive power of heat; with numerical results deduced from Regnault's experiments on steam', Trans. Royal Soc. Edinburgh, vol. 16 part 5, pp 541-574, (1849). Accessed at https://archive.org/details/transactionsofro16roy/page/541 and reprinted in Mathematical and Physical Papers, vol. 1, pp 113-155

(1882). Accessed at https://archive.org/details/mathematicaland01kelvgoog/page/n140

12. W.J. Maquorn Rankine, ' 'On the mechanical action of heat, especially in gases and vapours', Trans. Royal Soc. Edinburgh, vol. 20, part 1, pp 147- 193 (1853). Read on 4th Feb 1850. Accessed at https://archive.org/details/transactionsofro20royal/page/146

13. M.R. Clausius, 'On the moving force of heat, and the Laws regarding the nature of heat itself which are deducible therefrom', Philosophical Magazine ser. 4, vol. 2, no. 8, pp 1-20 (1851), reprinted and translated from the original (Feb 1850) paper in German in *Annalen der Physik*. Accessed at https://books.google.co.uk/books?id=JbwdWbbM1KgC&pg=RA1-PA1&redir_esc=y#v=onepage&q&f=false

14. William Thomson, 'On the dynamical theory of heat, with numerical results deduced from Mr. Joule's equivalent of a thermal unit, and M. Regnault's observations on steam', Trans. Royal Soc. of Edinburgh, vol. 20, part 2, pp 261-288 (1853). Read on 17th March 1851. Accessed at https://archive.org/details/transactionsofro20royal/page/260

15. David Saxon, 'In praise of Lord Kelvin', Physics World, (2007). Accessed at https://physicsworld.com/a/in-praise-of-lord-kelvin/

16. William H. Cropper, 'Great Physicists', pp 59-70 and 78-92, pub. Oxford University Press (2001)

17. Crosbie Smith and M. Norton Wise, Energy and Empire: A biographical study of Lord Kelvin', pub. Cambridge University Press (1989)

18. Picture credit Illustration 47: 'Lord Kelvin William Thomson (1824-1907), at the Age of 22' by John Graham-Gilbert. Glasgow Museums Resource Centre. Accessed at https://www.artuk.org/discover/artworks/lord-kelvin-william-thomson-18241907-at-the-age-of-22-84170

19. Picture credit Illustration 48: Lord Kelvin gravure. Accessed at https://commons.wikimedia.org/wiki/File:Lord_Kelvin_gravure.jpg under the GNU Free Documentation Licence v1.2 and Creative Commons Attribution-Share Alike 3.0 Unported licence.

40. The beautiful equations that changed the world

1. Basil Mahon, 'The Man who Changed Everything', pub. John Wiley & Sons (2004)
2. Lewis Campbell and William Garnett, 'The Life of James Clerk Maxwell', pub. Macmillan & Co. (1882). Digitised by James C. Rautio. Accessed at https://www.sonnetsoftware.com/bio/maxbio.pdf Reprinted by permission of Sonnet Software, Inc.
3. James Clerk Maxwell, W.D. Niven (ed.) 'The scientific papers of James Clerk Maxwell', pub. Dover (1965). Accessed at http://strangebeautiful.com/other-texts/maxwell-scientificpapers-vol-i-dover.pdf
4. Kevin Johnson, 'The life of James Clerk Maxwell', (2002). Accessed at http://mathshistory.st-andrews.ac.uk/Projects/Johnson/Chapters/Ch3.html
5. William H. Cropper, 'Great physicists', pp 154- 176, pub. Oxford University Press (2001). Accessed at Higher Intellect archive: https://cdn.preterhuman.net/texts/science_and_technology/physics/Great%20Physicists%20-%20From%20Galileo%20to%20Hawking%20-%20W.%20Cropper.pdf
6. Alexander Macfarlane, 'Lectures on ten British physicists of the nineteenth century', pub. John Wiley & Sons (1919). Accessed at https://archive.org/details/lecturesontenbri00macfrich
7. P.M Harman, 'Maxwell, James Clerk', Oxford Dictionary of National Biography (2009). Accessed at https://doi.org/10.1093/ref:odnb/5624
8. James Clerk Maxwell Foundation, 'The published scientific papers and books of James Clerk Maxwell (1831-79)'. Accessed at http://www.clerkmaxwellfoundation.org/published_scientific_papers.pdf (2020)
9. Picture credit Illustration 49: James Clerk Maxwell c1882, digitised from an engraving by G.J. Stodart from a photograph by Fergus of Greenock. Accessed at https://commons.wikimedia.org/wiki/File:James_Clerk_Maxwell_big.jpg {{PD-US-expired}}.
10. Picture credit Illustration 50: Calculated Maxwell-Boltzmann distribution of speeds for 1 million oxygen atoms at 3 temperatures.

Author Superborsuk. Accessed at https://commons.wikimedia.org/wiki/File:Maxwell-Boltzmann_distribution_1.png under GNU Free Documentation Licence v1

About the author

John Mellis has authored many technical papers and some not-so-technical articles in various journals and periodicals. This is his first book. He was born in Glasgow, where he studied Applied Physics, Logic and Semantics, and the Philosophy of Science at the University of Strathclyde. He gained a Ph.D. from the University of St Andrews, for experimental and computational research on the physics of CO_2 lasers. After a postdoctoral fellowship in high-power lasers funded by British Aerospace, he moved to England to work on optical signal processing at the Standard Telecommunication Laboratories (STL) in Essex. Most of his career has been with the BT Research Labs near Ipswich, working on optical communications networks and advanced software algorithms, on tech-based spinout ventures, and in global project management. For many years he was a Visiting Professor in the School of Computing and Technology at the University of Sunderland. He is a Fellow of the Institute of Engineering and Technology, and lives in Suffolk.

Printed in Great Britain
by Amazon